SO YOUR HOME IS BUILT ON EXPANSIVE SOILS

AF815378

A Discussion on How Expansive Soils Affect Buildings

Second Edition

Prepared by the Shallow Foundations Committee of the
Geo Institute of the American Society of Civil Engineers

Edited by
Warren K. Wray, Ph.D., P.E., D.GE., F.ASCE
Marshall B. Addison, Ph.D., P.E., M.ASCE
Kenneth M. Struzyk, P.E., M.ASCE

PUBLISHED BY THE AMERICAN SOCIETY OF CIVIL ENGINEERS

Library of Congress Cataloging-in-Publication Data

Names: Wray, Warren K., editor. | Addison, Marshall B., editor. | Struzyk, Kenneth M., editor.
Title: So your home is built on expansive soils / edited by Warren K. Wray, Ph.D., P.E., D.GE, Marshall B. Addison, Ph.D., P.E., Kenneth M. Struzyk, P.E.
Description: 2nd editon. | Reston, Virginia : American Society of Civil Engineers, [2019] | Includes bibliographical references and index.
Identifiers: LCCN 2019001765| ISBN 9780784415214 (print : alk. paper) | ISBN 9780784481950 (pdf) | ISBN 9780784481967 (epub)
Subjects: LCSH: Earth movements and building--Miscellanea. | Swelling soils--Miscellanea.
Classification: LCC TH1094 .S68 2019 | DDC 624.1/5136--dc23
LC record available at https://lccn.loc.gov/2019001765

Published by American Society of Civil Engineers
1801 Alexander Bell Drive
Reston, Virginia 20191-4382
www.asce.org/bookstore | ascelibrary.org

The material presented in this publication has been prepared in accordance with generally recognized engineering principles and practices and is for general information only. Any statements expressed in these materials are those of the individual authors and do not necessarily represent the views of ASCE, which takes no responsibility for any statement made herein. No reference made in this publication to any specific method, product, process, or service constitutes or implies an endorsement, recommendation, or warranty thereof by ASCE. The materials are for general information only and do not represent a standard of ASCE, nor are they intended as a reference in purchase specifications, contracts, regulations, statutes, or any other legal document. ASCE makes no representation or warranty of any kind, whether express or implied, concerning the accuracy, completeness, suitability, or utility of any information, apparatus, product, or process discussed in this publication, and assumes no liability therefor. The information contained in these materials should not be used without first securing competent advice with respect to its suitability for any general or specific application. Anyone utilizing such information assumes all liability arising from such use, including but not limited to infringement of any patent or patents.

ASCE and American Society of Civil Engineers—Registered in U.S. Patent and Trademark Office.

Photocopies and permissions. Permission to photocopy or reproduce material from ASCE publications can be requested by sending an email to permissions@asce.org or by locating a title in the ASCE Library (https://ascelibrary.org) and using the "Permissions" link.

Errata: Errata, if any, can be found at https://doi.org/10.1061/9780784415214.

Illustrations © by John Nunnemacher

Copyright © 2019 by the American Society of Civil Engineers.
All Rights Reserved.
ISBN 978-0-7844-1521-4 (print)
ISBN 978-0-7844-8195-0 (PDF)

Manufactured in the United States of America.

25 24 23 22 21 20 19 2 3 4 5

Contents

Foreword

The first edition of *So Your Home is Built on Expansive Soils* was commonly referred to as the "Yellow Book" because of the color of its cover. The purpose of the first edition was to help homeowners understand why expansive soils shrink and heave. In addition to homeowners, this book has been found to be a valuable resource for builders and realtors. Because of the overwhelming popularity of the original edition over the last couple of decades, there has been ongoing interest in producing a second edition. As a result, the Shallow Foundations Committee formed a subcommittee dedicated to the second edition update, which consists of the following members:

Warren K. Wray, Ph.D., P.E., D.GE., F.ASCE
Marshall B. Addison, Ph.D., P.E., M.ASCE

Kenneth M. Struzyk, P.E., M.ASCE
Derrick D. Dasenbrock, P.E., D.GE, F.ASCE

The second edition provides numerous updates to the book's content and illustrations. Sincere appreciation is given to Warren K. Wray for his leadership, Marshall Addison for his technical narrative updates, and Ken Struzyk (AccuTech Consultants LLC) for the new illustrations and photographs. The Shallow Foundations Committee welcomes comments on how information on expansive soils might be better presented and invites suggestions for additional items that might be included in any future editions of the book.

Derrick D. Dasenbrock, P.E., D.GE, F.ASCE
Chair, Shallow Foundations Committee

Preface

The problems associated with expansive soils are widely experienced in several areas throughout the United States. These soils can cause a great deal of damage to residential structures. *So Your Home is Built on Expansive Soils*, originally published in 1995, was intended for use by individuals dealing with homes that are supported on expansive soil sites. Initially edited by Dr. Warren K. Wray, this extremely popular document has been a valuable source of information for homeowners, builders, and realtors. Fondly referred to as the "Yellow Book," this publication has provided valuable insight with respect to shrink-swell soils.

As practicing engineering consultants, we have found the Yellow Book to be an easy-to-read reference for understanding the basic principles of expansive soils and their effects on foundation performance. As a result, we have referred the book to hundreds of clients. By understanding expansive soil behavior, individuals can implement changes that may reduce differential foundation movements and the resulting distress. These changes are primarily associated with moisture control measures that directly influence a clay soil's shrink-swell potential. Although the term *house* is usually used in the text, the information in this book is equally applicable to nonresidential low-rise buildings or structures.

This book is divided into two sections to allow for easier comprehension. Part 1 presents the basic principles of expansive soils and provides various procedures to reduce the shrink-swell potential of clay soils. Part 2 addresses common questions associated with foundations supported on expansive soils. This section of the book presents the information in a question-and-answer format. In the second edition, considerable new material is being presented, including narrative updates and additional illustrations we trust provide an enhanced understanding of expansive soils.

Marshall B. Addison, Ph.D., P.E., M.ASCE
Kenneth M. Struzyk, P.E., M.ASCE

Acknowledgments

Arizona State professor of unsaturated geotechnics, Sandra Houston, remarks "Kent Wray has provided key research into the development of sound suction-based methods for estimation of soil heave. He has provided leadership for clear communication with builders, contractors, and homeowners on what causes clays to shrink and swell, and on best practices from pre-construction through construction to post-construction occupancy and use."

Sandra Houston is a past president of the Geo-Institute. She spearheaded the formation of and served as the first chair of the GI Committee on Unsaturated Soils.

***

Gordon McKeen, long-time colleague of Dr. Wray's, shares "Kent Wray's work provided practicing engineers with tools for educating clients, owners, and anyone involved with shallow foundations. The original work was a concise presentation of technical issues as well as a guide to the application of common sense essential to adequately address the issues. I applaud his additional contributions to the second edition as a great service to the practicing engineer community."

Gordon McKeen is a practicing engineer in Albuquerque, New Mexico; he retired in 2002 as a research program manager at the University of New Mexico.

***

As with the earlier book, Marshall Addison and Kenneth Struzyk state "We are indebted to Dr. Wray for his contributions to the publication of this edition. His knowledge of shrink-swell soils and his practical advice to homeowners and consultants has made living with these problematic soils less stressful. Therefore, we are extremely honored and grateful to have worked with Dr. Wray and taken a part in the preparation of this second edition.

In addition, the subcommittee expresses their thanks to Caleb Douglas, Philip King, and Daniel Rich for their early contributions that led to the preparation of the second edition of the book.

Acknowledgments

Arizona State professor of unsaturated geotechnics, Sandra Houston, remarks "Kent Wray has provided key research into the development of sound suction-based methods for estimation of soil heave. He has provided leadership for clear communication with builders, contractors, and homeowners on what causes clays to shrink and swell, and on best practices from pre-construction through construction to post-construction occupancy and use."

Sandra Houston is a past president of the Geo-Institute. She spearheaded the formation of and served as the first chair of the GI Committee on Unsaturated Soils.

***

Gordon McKeen, long-time colleague of Dr. Wray's, shares "Kent Wray's work provided practicing engineers with tools for educating clients, owners, and anyone involved with shallow foundations. The original work was a concise presentation of technical issues as well as a guide to the application of common sense essential to adequately address the issues. I applaud his additional contributions to the second edition as a great service to the practicing engineer community."

Gordon McKeen is a practicing engineer in Albuquerque, New Mexico; he retired in 2002 as a research program manager at the University of New Mexico.

***

As with the earlier book, Marshall Addison and Kenneth Struzyk state "We are indebted to Dr. Wray for his contributions to the publication of this edition. His knowledge of shrink-swell soils and his practical advice to homeowners and consultants has made living with these problematic soils less stressful. Therefore, we are extremely honored and grateful to have worked with Dr. Wray and taken a part in the preparation of this second edition.

In addition, the subcommittee expresses their thanks to Caleb Douglas, Philip King, and Daniel Rich for their early contributions that led to the preparation of the second edition of the book.

Part I
A Discussion of How Expansive Soils Affect Buildings

Introduction

Soils with the potential to shrink or swell are found throughout the United States and in almost all parts of the world (Figure 1-1). Soils with this shrink-swell potential create difficult performance problems for buildings constructed on these soils because as the soil water content increases, the soil swells and heaves upward, and as the soil water content decreases, the soil shrinks, and the ground surface recedes and pulls away from the foundation.

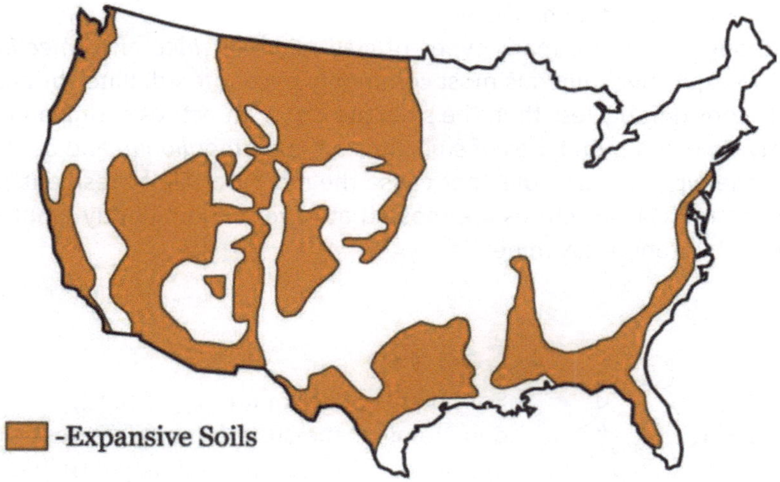

Figure 1-1. Expansive soils map of the United States.
Source: AccuTech Consultants, LLC, used with permission.

The effect of expansive soil damage on a local, regional, or national scale is considerable. Among the first persons to attempt to quantify the extent of damage resulting from expansive soil movement were Jones and Holtz (1973), who estimated the annual cost of expansive soil damage in the United States at an estimated $2.2 billion and to exceed that caused by earthquakes, hurricanes, and floods combined in an average year. Krohn and Slosson (1980) estimated the annual cost of expansive soil damage in the United States to be $7.0 billion in 1980. Krohn and Slosson further estimated that damages to single-family and commercial buildings accounted for nearly one-third of the total amount of damage resulting from expansive soils. An earlier damage survey conducted solely in Dallas County, Texas, identified 8,470 residential foundation failures that occurred in only one year (1974), 98% of which occurred in expansive soils (Wray 1989).

Expansive Soils

Expansive soils are also known as *swelling*, *heaving*, and *shrink-swell* soils. In the United Kingdom, these soils are known as shrinkable soils. By whatever name they are called, expansive soils are clay soils. Sometimes the clay has been compressed by great weight at some time in its geologic past and is called *shale*, which can also be expansive. Nearly all clay soils swell when they get wetter and shrink when they get drier. Those that shrink and swell to extremes are called *expansive* soils.

Although there are many types of clay minerals, *illite* and *smectite* are the two expansive minerals most commonly encountered. Illite shrinks and swells considerably less than the smectite clays. Smectites shrink and swell greatly. Two common types of smectites are *montmorillonite* and *bentonite*. Thus, the types of clay soils that cause the most trouble to residential and light commercial structures are those that have predominantly montmorillonite or bentonite clay minerals.

Soil Tip

Soils that change volume significantly, depending on moisture content, are described by many different names, often determined by geographic region or country. Although commonly called *expansive soils* in the United States, these soils both expand and contract.

Shrink or Swell

As clay particles are formed, there are usu... e
arrangement where there is an electrical imba...
is increased whenever a "string" of clay particl...
particle typically has a negative net electrical c...
nature likes all things to be balanced, whenever ...
enough to the surface of a clay particle, the negativ...
clay particle causes the positive end of the water mole...
particle and, if close enough to the particle, the water ...
to the clay particle surface strong enough that the water ...
trapped. Also, unattached (free) positively charged particle...
tend to acquire a spherical-shaped arrangement of water mo...
have their negative ends directed toward the positively charged ... nd
their positive ends directed away from the cation). When the free c... on and
its captured water molecules approach a clay particle, the attraction between the negatively charged clay particle surface and the positively charged outside of the cation's sphere of water molecules causes the cation to be captured by the clay particle, thus increasing the amount of water associated with the clay particle (Figure 1-2).

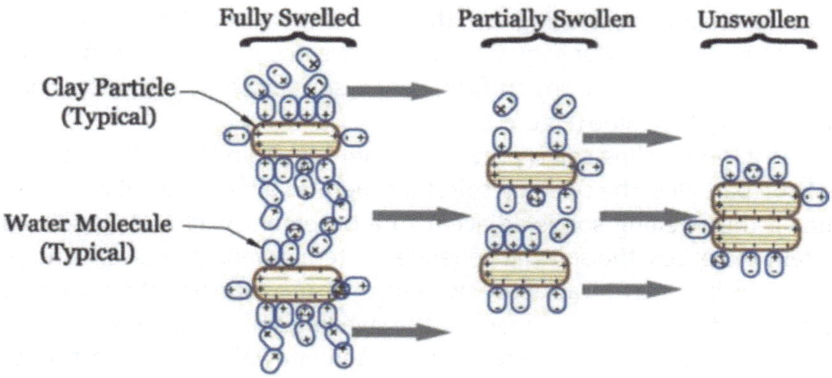

Figure 1-2. Clay particle water transfer: arrows represent the direction of movement of water and cations toward clay particles with higher electrical imbalance.

Source: AccuTech Consultants, LLC, used with permission

Clay particles are very small. A typical nonexpansive clay particle might have a total surface area (top, bottom, and edges) of approximately 1×10^{-5} mm^2 (1×10^{-10} ft^2, or 0.0000000001 ft^2). As areas go, this is very small! Smectite particles have a diameter that is 100 to 1,000 times smaller than the most nonexpansive clay and a thickness that is 10 to 400 times thinner (Grim 1968) and, consequently, typically have a larger surface area per particle. Thus, a single pound (0.45 kg) of montmorillonite particles would have an incredible total surface area of approximately 800 ac (325 ha) (Dixon and Weed 1977) with which to attract water!

Thus, expansive soils are very small in size and have a large surface area that attracts free water. Because of these characteristics, expansive soils are those clays that exhibit an extreme change in volume.

Heave. When the term *heave* is used with respect to expansive soils particles, it usually means that the soil surface is moving upward. However, if given the opportunity to do so, an expansive soil that is getting wetter will increase in volume, or heave, in every direction. However, because the expansive soil particle adjacent to the particle that is attempting to expand laterally is itself attempting to expand laterally, the result is rather like the soil engaging in an isometric exercise: it is simply pushing against itself and, because of that, it cannot expand sideways. Consequently, expansive soils near the surface most often expand upward (i.e., heave).

For the soil to heave or expand upward, it must push the soil above it upward, too. Some swelling soils have been measured to exert many tons per square foot of swelling pressure. If the heaving soil is near the surface, it is very easy for the underlying swelling soil to push up the soil above it. However, the deeper the swelling soil, the greater the amount of soil above that must be pushed up. Consequently, less heave occurs deeper in the soil and more heave occurs near the surface; hence, soils closer to the surface have a greater weighted value (Figure 1-3).

Even if the swelling soil is prevented from expanding laterally by the adjacent soil particles, the swelling soil still generates a lateral swelling pressure. Thus, if the swelling soil is adjacent to a basement wall and its soil water content increases, the soil will generate a lateral swelling pressure and push against the basement wall. If the swelling pressures are exceptionally large or if the soil is too densely compacted, the swelling soil can damage the wall. If the wall is a *retaining* wall rather than a basement wall, which is restrained at the top, the swelling soil can push the retaining wall outward, and even push it over if the heaving is severe enough.

Conversely, if the soil is drying out, the soil will also change in volume in every direction but in a direction opposite to that which occurs when it is getting wet (i.e., the soil shrinks in volume). When the soil shrinks, one soil particle does not resist shrinkage by an adjacent soil particle like it resisted swelling by the same particle. Instead, every soil particle is free to reduce in size by giving up water. Besides causing the ground surface to recede or "go down," the shrinking process results in cracks in the soil.

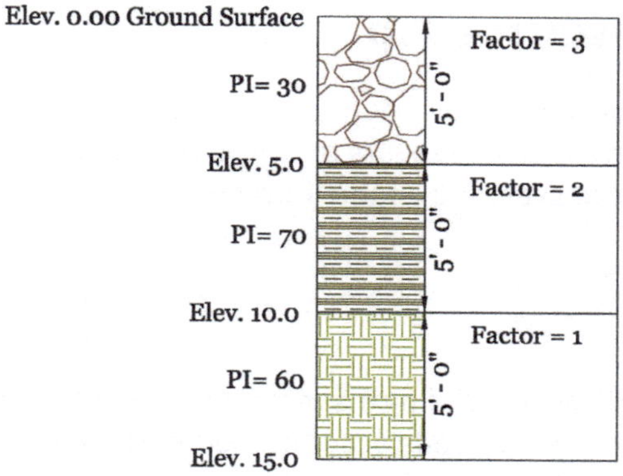

F	D	FxD	PI		FxDxPI
3	5'-0"	15	30	=	450
2	5'-0"	10	70	=	700
1	5'-0"	5	60	=	300
	Total =	30			1450

Equivalent P.I. = 1450/30 = 48.33
= 48

Figure 1-3. Soil near the surface has three times more influence on heave motions than deeper soil.

Source: AccuTech Consultants, LLC, used with permission.

In short, expansive soils heave in all directions when the soil particles acquire additional water, but the upward movement of the surface becomes significant when the lateral expansion of the wetter clay particles is prevented from swelling any more in the horizontal direction. Thus, the soil continues to expand in the vertically upward direction. When the soil begins to dry out, the soil particles shrink in all directions, creating cracks in the soil, and causing the ground surface to recede downward.

How to Determine if Your Home is Built on Expansive Soil

The best and most definitive way to identify the type of soil beneath your house is to ask a geotechnical engineer to determine whether the soil is expansive. However, there are other ways of determining if the soil is expansive besides engaging a geotechnical engineer.

By entering an address into the USDA Natural Resources Conservation Service (NRCS) online Web Soil Survey, an aerial photograph and map of the area may be viewed. The maps include a printed number or letter code that identifies a soil type. A short, written description of each soil type can be found at that website and also several tables of engineering and applications information are given in the tables section of the report. The information in the text and tables will identify the shrink-swell tendencies of each soil type.

Resource Tip

A quick method of determining if your house is on expansive soils is to view a county soils report, available online from the NRCS.

City or county engineering offices or building inspection offices often have a general knowledge of where expansive soils occur in their particular jurisdictions. A visit to either of these offices, which should also be free (although there might be a nominal charge if any documents are reproduced for you), may be sufficient to determine the type of soil your home or business is built on.

The surest way to determine if your site has expansive soils is to engage a geotechnical engineer. The engineer will make one or more soil borings on your site and extract soil samples from below the ground surface. The extracted samples will be taken to a soils or geotechnical engineering laboratory and tested. The test results will show if the soil is expansive and to what depth and, to some degree, just how expansive the soil is if expansive soil is indicated. The engineer should provide you with a written report that is sealed by a registered professional engineer. This method of learning whether your site contains expansive soils is considerably more accurate than either of the aforementioned methods, but it will be more expensive.

How to Solve the Problem

There are two conditions that must be satisfied before expansive soils become a problem: expansive soils must be present, and the soil moisture condition must change.

Obviously, if expansive soils are not present, the extreme soil shrink and heave normally associated with expansive soils will not occur.

If the soil water content can be kept from changing, or at least the change kept to a minimum, lesser shrink or heave will occur and the problem created by expansive soils will be minimized. However, constructing a house or a building interrupts an established energy gradient (principally because of surface evaporation and plant transpiration) that is causing soil water to move from depth to the surface or vice versa. This induced water flow will ultimately result in some shrink or heave, even in the volume of soil beneath the interior of the house or building that is not being influenced by outside factors such as climate. However, once the interrupted energy gradient has reached equilibrium, no further shrink or heave will likely occur unless something happens to change the soil moisture equilibrium. Three things that can often cause the soil water content to change are climate, site vegetation, and irrigation.

Engineer's Tip

Distress and damage can be reduced by taking steps to maintain the water content of the soil beneath and adjacent to your home. If your home is in a predominantly wet climate, the goal is to prevent the soil from drying out; if you live in a predominantly dry climate, the goal is to prevent the soil from getting wet. Problems occur as a result of changes in the moisture content and the associated effects on the volume of shrink-swell soils.

Climate. Expansive soils occurring in predominantly wet climates or predominantly arid climates typically do not produce the extent of damage that expansive soils occurring in semiarid climates cause. Expansive soils in predominantly wet climates have, for the most part, already acquired nearly all the soil moisture needed to produce soil heave. The damage to structures built over expansive soils in wet climates most often occurs during periods of drought. The United Kingdom is in a predominantly wet climate, and although many of the clays occurring in that country are some of the potentially most active expansive clays in the world, little damage occurs to structures in the United Kingdom except during periods of drought or periods of less than usual rainfall. This is likely the reason that expansive soils are termed *shrinkable soils* in the United Kingdom. In the United States, Houston, Texas, and Tulsa, Oklahoma, are located in relatively wet climates. Droughts occurring during the 1980s produced millions of dollars of total damage to residential and other lightly loaded structures in those two cities because the normally wet soils dried and shrank in volume.

Conversely, expansive soils that occur in arid climates typically do not cause much damage to structures constructed over them unless the clay experiences a major wetting period or episode. For example, houses in Amarillo, Texas, experienced considerable damage in the early 1980s when a part of the city constructed in an expansive soil region was flooded in August, the hottest and driest month of a long, hot, and dry summer that year.

Most damage to structures from expansive soil movement occurs in locations that have a *semiarid* climate, which can be described as a climate with periods of rainfall followed by long periods of no rainfall. This type of climate typically exhibits rainfall over a period of several weeks, which results in the soil becoming wetter and swelling. However, the rainy season is then followed by a longer period with little or no rainfall, during which the soil gives up the moisture that it acquired in the preceding rainy period, dries out, and shrinks. Houses supported on shallow foundations in a semiarid climate experience an annual cyclic rise and fall of the structure as the soil heaves, shrinks, and heaves again (Figure 1-4).

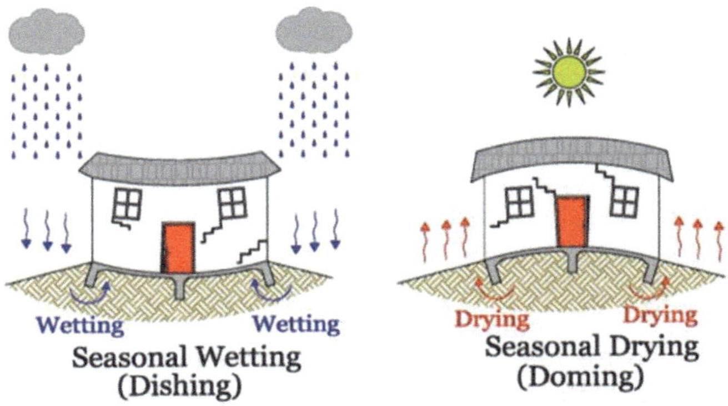

Figure 1-4. Seasonal moisture change.
Source: AccuTech Consultants, LLC, used with permission.

REMEMBER

Similarly, you should take steps to ensure that the soil water content of the soil beneath and adjacent to your home does not appreciably increase if you live in a predominantly dry climate.

The task of ensuring that the soil water content remains essentially constant becomes more challenging if you live in a semiarid climate, but this task is no more challenging to any conscientious homeowner than other tasks related to good lawn or home maintenance. Some of the homeowner tasks to be undertaken are discussed in a later section.

Thus, the solution to controlling soil movement due to climate is to ensure that the soil beneath and around your home does not dry out if your home is in a predominantly wet climate.

Landscaping Tip

Plan your plants! A small sapling will grow into a much larger tree over time. Make sure trees and shrubs are planted well away from your foundations to help maintain consistent levels of moisture. Planting them further away, or selecting vegetation with minimal water needs, is an even better alternative to help minimize changes in moisture near the foundation.

Vegetation. Vegetation has been shown to be an important factor in causing changed soil moisture conditions. Vegetation can affect the soil conditions around and under a house or other structure in several ways. A common practice for watering shrubs, bushes, and flowers planted adjacent to a house or other building in the southwestern United States is to excavate the vegetation bed a few inches below ground level and then let a garden hose run water into the depression until the bed is filled with water. The frequent result of this flooding practice is to induce water flow beneath the house or structure with a subsequent heaving along the perimeter of the building. Obviously, this is not the fault of the vegetation, but rather the fault of the owner for engaging in a poor watering practice.

Often, the vegetation itself can produce damage to structures. The damage is most frequently the result of plants withdrawing water from the soil and causing the soil to dry out and shrink. Large bushes or shrubs planted immediately along the outside of a building can withdraw water from under the edge of the building if the plants are not watered regularly. Trees are more frequently the cause of this type of plant-induced soil shrinkage. Numerous instances of severe damage to buildings have been documented where roots from nearby trees have penetrated beneath a building's foundation and removed water from the soil beneath the building during periods of drought (Figure 1-5).

The recommended practice for planting trees is to plant them far enough away from the building so that roots will not grow back underneath the building. This recommendation is usually hard for owners to follow because owners like to have large, shady trees around their houses or buildings. In general, roots grow out a little beyond the edge of the tree's limbs so that the roots can extract water entering the soil from the ground surface during rains or landscape irrigation (Figure 1-6).

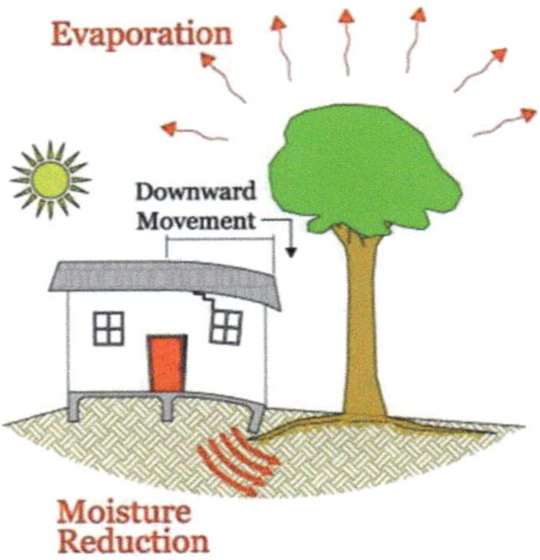

Figure 1-5. Tree effect.

Source: AccuTech Consultants, LLC, used with permission.

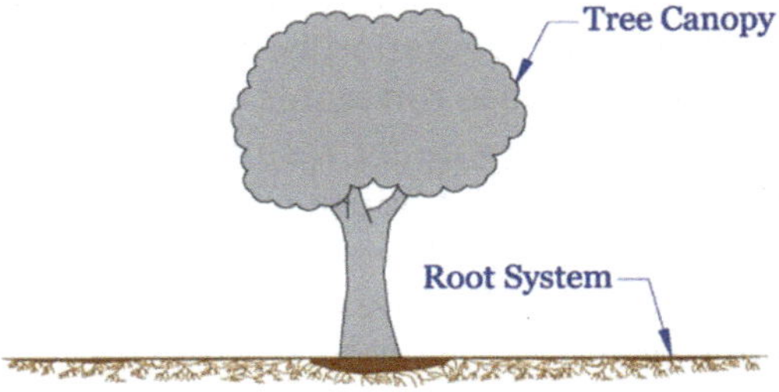

Figure 1-6. Tree root location versus canopy location.

Source: AccuTech Consultants, LLC, used with permission.

This edge of the tree's limbs and leaves is sometimes referred to as the *drip line*.

REMEMBER

One rule of thumb that has been shown to be successful is to plant trees no closer to the building than where a mature tree's drip line will be 20 years after it is planted as a sapling.

Another rule of thumb that is also used, but is a little more conservative, is to plant the tree a distance away from the building equal to the mature height of the tree (Figure 1-7).

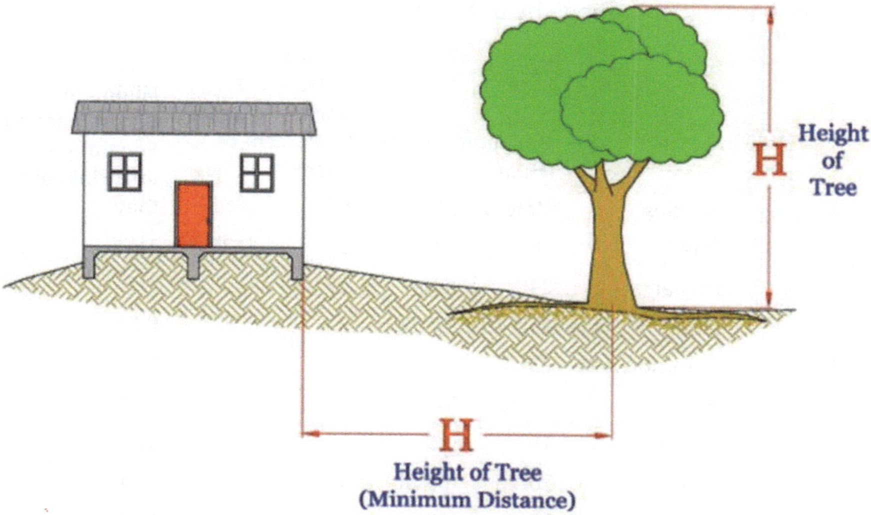

Figure 1-7. Tree planting location.

Source: AccuTech Consultants, LLC, used with permission.

Rules of (a Green) Thumb

The overall landscaping goal is to ensure that the root system from plants and trees is kept outside of the foundation area so that the soil moisture changes are limited. There are several different rules of thumb, but each tries to ensure an appropriate offset between the vegetation and the foundation.

Arborists and nursery persons can usually tell about how tall a tree will grow at maturity. The UK Royal Botanic Gardens suggests that the trees most likely to cause damage, in descending order of threat, are those shown in Table 1-1.

Table 1-1. Risk of Damage by Different Varieties of Tree.

Ranking	Species	Maximum Height (H) of Tree,		Separation Between Tree and Building for 75 Percent of Cases,		Minimum Recommended Separation In Shrinkable Clay
		Meters	(Ft)	Meters	(Ft)	
1	Oak	16-23	(50-75)	13	(43)	1H
2	Poplar	24	(80)	15	(50)	1H
3	Lime	16-24	(50-80)	8	(25)	0.5H
4	Common Ash	23	(75)	10	(30)	0.5H
5	Plane	25-30	(80-100)	7.5	(25)	0.5H
6	Willow	15	(50)	11	(35)	1H
7	Elm	20-25	(65-80)	12	(40)	0.5H
8	Hawthorn	10	(30)	7	(23)	0.5H
9	Maple Sycamore	17-24	(55-80)	9	(30)	0.5H
10	Cherry/Plum	8	(25)	6	(20)	1H
11	Beech	20	(65)	9	(30)	0.5H
12	Birch	12-14	(25-45)	7	(23)	0.5H
13	White Beam	8-12	(25)	7	(23)	1H
14	Rowan	8-12	(25-40)	7	(23)	1H
15	Cypress	18-25	(60-80)	3.5	(10)	0.5H

Note: Adapted from BRE Digest (1985).

Vegetation can result in structural damage to new buildings on sites where vegetation was removed shortly before construction. When constructing new homes or buildings that cover large areas, it is often the practice to remove all trees and large shrubs at the time the site is being leveled and graded. If trees or large shrubs are removed at the end of the dry season or at the end of a drought during this construction operation, the ground beneath and around the trees and shrubs will most likely be very dry and even desiccated. If the building is subsequently built over, or adjacent to, the desiccated site, the soil will subsequently rehydrate (*wet up*) once the ground surface is covered (Figure 1-8). If artificial irrigation is also used at the site, the resulting post-construction heaving is exacerbated, and the heaving that occurs will likely be more than would have occurred without the irrigation.

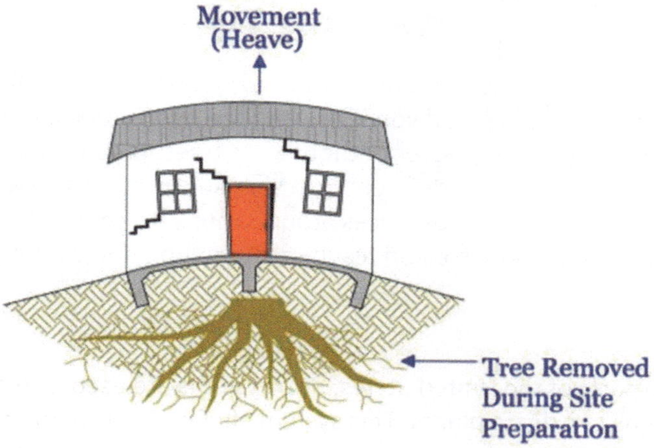

Figure 1-8. Rehydration of formerly drier soils and resulting heave due to construction over recently removed tree.

Source: AccuTech Consultants, LLC, used with permission.

Thus, shrinking or heaving resulting from vegetation-related problems can also be controlled by simply understanding the effect vegetation has on soil water content and the impact of the location of the vegetation on the performance of the building. Shrinking and heaving can also be controlled by understanding how watering of vegetation affects soil water content and the subsequent soil movement.

Landscape Irrigation. Most people like to have lush, green lawns around their houses and often like to have similar attractive entrances to their businesses or other buildings. Often an automatic or semiautomatic sprinkling or irrigation system is installed to ensure that the lawn and bushes, shrubs, and flowers receive regular watering. Owners may think that if, for example, 1 in. (25 mm) of irrigation water twice a week is good for the lawn, then greater amounts or more frequent watering of the same amount is even better. Although the lawn and plants may not object to the additional watering, all too often the result is an increase in soil water content around the edge of the house or building. The increased soil water content, in turn, causes the soil around the edge of the building to heave, resulting in building damage. Thus, when automatic sprinkling systems are involved, it is best to water only enough to satisfy the water demands of the lawn and the plants on the property, including the trees.

Watering Tip

It is important to understand the watering needs of your lawn and the plants on your property. Even without large trees, which contribute to differences in soil moisture, uneven watering practices around your yard could lead to local shrinking or heaving which could result in foundation movement and distress.

In some parts of the United States, drip systems are used instead of sprinkling systems to irrigate plants. This system allows water to drip or seep into the soil at the root level. A drip system is often preferable to a conventional sprinkling system because it provides water at the point where it is needed without saturating the surface. Drip systems are commonly used in California and other parts of the Southwest to water vegetation planted adjacent to the house. Drip systems are more frequently being required by city ordinances in regions prone to droughts and with scarce water resources.

Soil shrink and swell problems associated with underwatering as well as overwatering lawns and other vegetation can be reduced by controlled irrigation practices (Figure 1-9).

REMEMBER

It is important to understand the watering needs of your lawn and the plants on your property and to address those needs in a timely and conscientious fashion.

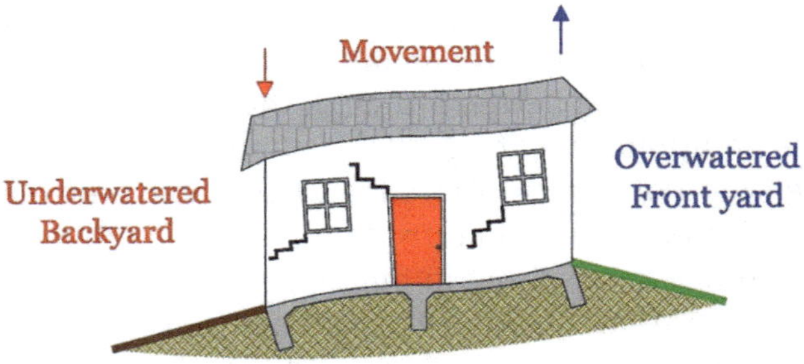

Figure 1-9. Improper watering effect.
Source: AccuTech Consultants, LLC, used with permission.

Foundation Types

There are two principal types of foundations used to support houses and other lightly loaded buildings that are constructed without a basement: *slab-on-grade* and *post-and-beam*. Houses without basements will be addressed first.

Engineer's Tip

Homes are built on a variety of foundation types, and some homes may even have more than one type of foundation, such as a home with a basement under most of the house and a slab foundation for an attached garage. Typically, homes will be built using floor slabs, posts and beams, basements, drilled piers, or combinations of these systems for support.

Houses without Basements. Although houses or other lightly loaded structures without basements can be built nearly anywhere, most buildings with basements are constructed in northern climates. Basements are not required in southern climates because the depth to which freezing temperatures reach in the soil is very shallow, therefore most structures in the South and Southwest are constructed with either slab-on-grade or post-and-beam foundations. The difference between the two types of foundations is that the slab-on-grade (often called *slab-on-ground*) foundation rests directly on the underlying soil. A post- or pier-and-beam foundation has a shallow void or *crawl space* underneath the building because the floor is supported on joists. The joists, in turn, are supported on cross beams. Cross beams are supported by short columns or piers, which ultimately transmit the structural load to the foundation. Post-and-beam foundation usually consists of a shallow, small diameter, usually concrete, pad or block that is typically buried at least 18 in. (450 mm) into the natural soil.

- Slab-on-grade (or just slab, for short) foundations usually are constructed of reinforced concrete and can be inexpensive depending on the price of concrete at time of construction. The function of a slab-on-grade foundation is not to resist or limit the amount of heave that might occur beneath the slab foundation, but to move up and down with the shrink and heave that might occur and to limit the distortion that the shrinking and heaving soil might cause to occur in the superstructure of the house. The amount of deflection or distortion that the slab is designed to permit is usually a function of the materials that will be used in the construction of the building being supported by the slab and the materials used to finish the structure (i.e., provide the finished appearance). Thus, *total* shrink or heave is usually less a design consideration for slab-on-grade foundations than the maximum expected *differential* deflection.

Slabs-on-grade are usually of two types: thin stiffened slabs (Figure 1-10) or uniformly thick slabs with thickened edges (Figure 1-11). The thin stiffened slab characteristically has a slab floor 4 or 5 in. (100 or 125 mm) thick with approximately 12 in. (300 mm) wide beams extending below the thin slab and typically spaced 10 to 20 ft (3 to 6 m) apart in both directions. These beams, called *grade beams*, typically range in depth from 18 to 30 in. (460 to 760 mm), measured from the top of the slab to the bottom of the beam (Figure 1-10). The beam (usually continuous) that goes around the outside of the foundation is called the *exterior* or *perimeter* grade beam, whereas those inside the perimeter grade beam are usually called the *interior* grade beams. It is important that the grade beams extend continuously across the cross section of the slab, that is, the grade beams should extend continuously from the left perimeter grade beam to the right perimeter grade beam to ensure the rigidity of the slab structure.

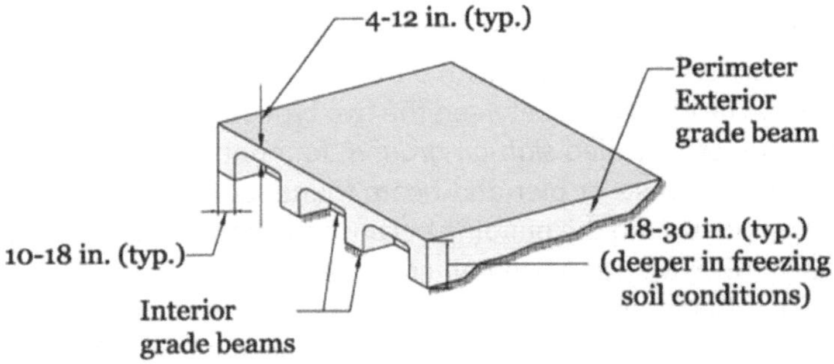

Figure 1-10. Sketch of typical stiffened slab-on-grade foundation.
Source: AccuTech Consultants, LLC, used with permission.

The uniformly thick slab typically is 4 to 12 in. (100 to 300 mm) thick with a thickened edge forming an exterior or perimeter grade beam (sometimes called a *perimeter footing*) that is 12 in. to sometimes as much as 30 in. (300 to 760 mm) thick. The uniformly thick slab is used more frequently in California, and the thin stiffened slab is used more frequently in other parts of the South and Southwest (see Figure 1-10).

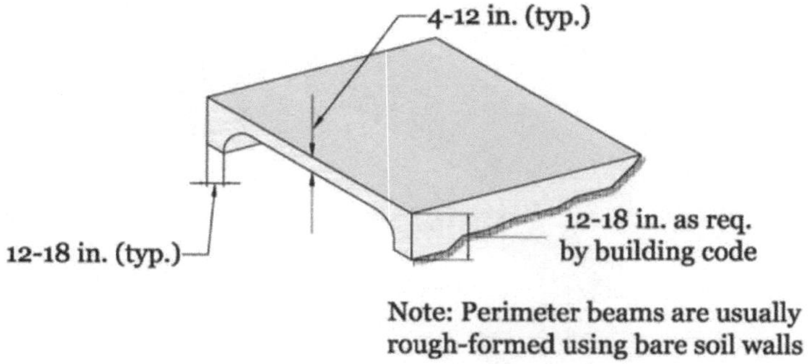

Figure 1-11. Typical uniformly thick slab with thickened edges.
Source: AccuTech Consultants, LLC, used with permission.

Typically, there are three types of reinforcement used in slab foundations: welded wire fabric (WWF, or just wire), mild steel (steel bars, or "rebars"), and post-tensioning (PT). None of the three types of reinforcement make any significant contribution to the strength of the concrete. The purpose of the reinforcement in slab foundations is principally to control cracking that will occur early in the life of the slab as the concrete sets and becomes hard. Welded wire fabric is not usually recommended for slab foundations to be constructed over expansive soils unless WWF sheets supported on reinforcement chairs are used. Mild steel-reinforced slabs require deeper beams or thicker uniform slabs than post-tensioned slabs. The reason for this is that post-tensioning induces precompression into the concrete slab, which permits the slab to withstand more tensile stress than the mild steel-reinforced slab before cracking.

In some parts of the country, only the perimeter grade beam or footing is reinforced, and the slab has no reinforcement other than light welded wire fabric or *fibermesh* (small, short, hairlike filaments mixed with the concrete) for crack control. These types of slab-on-ground foundations normally are not recommended for expansive clay areas.

- Post-and-beam foundations (Figure 1-12) provide a little more flexibility in the superstructure of the building that is supported on the foundation. They also provide an access to make repairs to utility lines under the house. Repairs to utility lines beneath a post-and-beam house are usually less expensive than if the same repairs are made to utility lines beneath both slab and post-and-beam foundations. If differential movement occurs between different parts of the house or building, it is possible to relevel the structure by placing jacks on the affected piers and jacking the structure back to level.

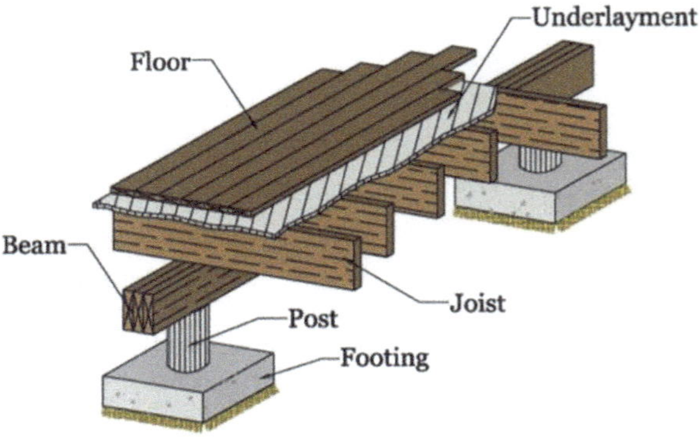

Figure 1-12. Typical post-and-beam foundation.

Source: AccuTech Consultants, LLC, used with permission.

Houses with Basements. Houses with basements are most often constructed in northern climates because the depth to which freezing temperatures advance into the soil during the winter requires the bearing element of the foundation to be at a considerable depth—often 4 ft (1.2 m) or more—below the ground surface. Thus, it is often convenient to simply excavate a basement and construct basement walls that perform the same function as a deep grade beam or deep footing to transmit the structural load to the supporting soil. The basement also has the additional advantage of increasing the living space of the house. The advantage that a basement presents to a house constructed in expansive soil is that the bottom of the basement is typically well below the depth over which the majority of the soil's shrink or swell due to climate occurs. However, basements present two additional problems that are not encountered when surface-supported foundations are used: basement floor heave and lateral wall heave (Figure 1-13).

Foundations with Piers. Sometimes a house is a slab-on-grade foundation, but the slab is supported by drilled piers. Drilled piers are long, slender, cylindrical columns of concrete that carry the weight of the structure being supported by the slab foundation to a deeper depth. Sometimes the piers are *belled* or *underreamed*, which means that they have a base that is wider than the column (or stem). The purpose of the bell or underream is to spread out the load over a wider area and reduce the increased pressure that the soil is experiencing from the building's weight or it reduces the amount of settlement that might occur from the increased stress. Often, however, the objective of the pier is to concentrate the structure's load to counteract the uplift pressure that might occur from the swelling soil. If the objective is to increase the pier loading, then the pier is constructed without a bell or underream, and the pier is called a *straight pier*.

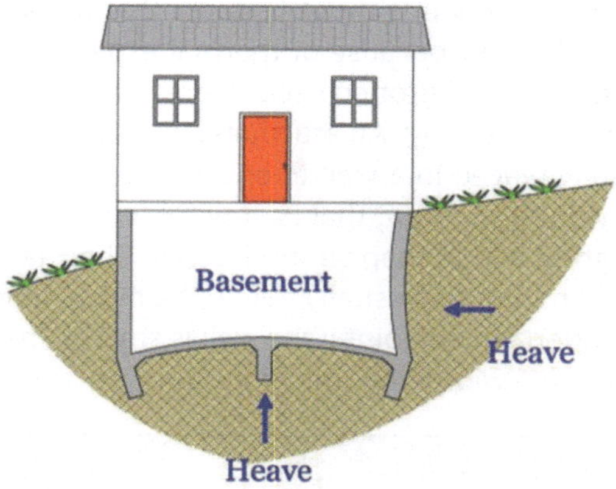

Figure 1-13. Cross section of typical basement foundation.
Source: AccuTech Consultants, LLC, used with permission.

If the house has a basement, piers can also be used to support the basement walls. When the basement walls are supported on piers, the basement floor is usually structurally separated from the basement walls. Sometimes the expected total amount of heave beneath the foundation is so great that it exceeds the capacity of a slab-on-grade to perform its job correctly. Most structural and geotechnical engineers consider 4 in. (100 mm) as the maximum amount of swell that should be permitted beneath slab-on-grade foundations. In these instances of excessive heaving, a *structural floor slab* is used. A structural floor slab performs just like a floor in a multistory building: it is intended to structurally support itself without any assistance from the underlying soil. A structural floor slab can provide as much clear space between the soil surface and the bottom of the floor slab as desired. The soil beneath the slab is then free to shrink and heave without affecting the slab performance. The structural slab is usually supported by perimeter beams (if the floor plan is large, there may also be interior beams which connect to the perimeter beams) which, in turn, typically are supported by drilled piers which carry the structure's weight down to a depth below the soil's active zone.

Causes of House Cracks

Cracking in the exterior of a house or building may occur for different reasons than those that occur in the structure's interior.

Exterior Cracks. Cracks are more readily noticed if the building exterior is brick, rock, or stucco. These types of finishes are brittle and susceptible to several different types of stresses.

Often, if a wall is long and without features that break up the length of the wall, a crack will occur because of thermal expansion and contraction. An architectural feature that is often used to prevent thermal stress cracks is called *articulation* (Figure 1-14). An articulated wall is one in which a vertical joint is purposefully placed in a wall to ensure that if a crack occurs, it will occur where we want it to occur, that is, it will occur at the articulation line and be less unsightly and perhaps unnoticeable. Thermal stress cracks are typically vertical in nature and usually follow the mortar line between bricks and rocks or an underlayment joint if the wall is stucco. Thermal stresses are usually about the same width from top to bottom.

Figure 1-14. Constructed joint to create articulated walls.
Source: Marshall B. Addison, Ph.D., P.E., used with permission.

Fill settlement cracks are often termed *stair step* (or in-echelon) cracks; this name results from the crack stair stepping from one course of bricks to the next through the mortar joints (Figure 1-15). If time is not a crucial element, it is often easy to distinguish between settlement cracks and shrink/heave cracks by observing the width of the crack as the climate changes from season to season. If the crack width changes with the change in season, the crack is the result of shrink and swell, not settlement. An observant homeowner in Beaumont, Texas, measured the width of a crack in his house as an indicator of how dry his lawn was getting during the summer months. He noted that the crack at the upper corner of his garage door got wider when the lawn needed watering and that the crack got smaller after watering. He also noticed that the crack began to close as summer turned to autumn and narrowed during the cooler, wetter winter months, but began to widen again as the months entered the hotter, drier summer season.

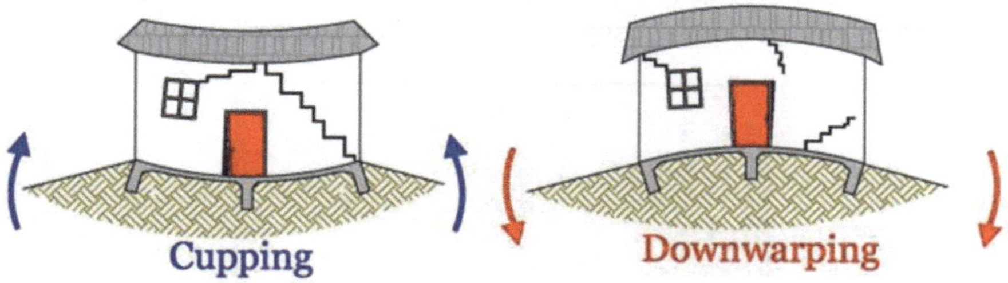

Figure 1-15. Heave- and shrinkage-induced stair step cracking.
Source: AccuTech Consultants, LLC, used with permission.

Take a Look Around

Cracks have various causes, and some are unavoidable as homes age and experience seasonal cycles. Heave and shrinkage stair step cracking is usually easy to spot because it starts at corners of doors or windows or at the edges of exterior walls.

Most engineers will recommend two things before repairs are begun to a house or building that has been damaged by expansive soil movement. The first recommendation is that the building's ground floor be surveyed to ensure that the type and degree of deformity or distortion is identified. Most floor elevation surveys are conducted using short focal distance survey leveling equipment, including lasers, digital gas manometers or traditional water levels; the cost of these services is typically several hundred dollars, depending on the location, the extent of the survey, the floor plan and/or floor area, the floor plan features and the ease with which the floor plan can be surveyed, and the distance the engineer has to travel.

The second recommendation is that a period of time be permitted to elapse during which the house is observed so that the range of movement or distortion is measured and observed as well as to ensure that the movement is neither continuing nor increasing. A one-year observation period is often recommended when structural failure is not a consideration. It makes no sense to rush into a repair when continuing movement can render the repair ineffective.

Stair step and diagonal cracks often begin at the corner of a door or window. They can also be associated with movement of an exterior wall. Although sometimes they can be related to thermal stress cracking, stair step cracks are almost always the result of either settlement or shrink-swell soil movement. Another indicator of corner movement is an opening of the facia joint at the corner of the house or separations in siding (Figure 1-16).

Figure 1-16. Siding separation.

Source: Marshall B. Addison, Ph.D., P.E., used with permission.

Houses with slab foundations are affected by the climate and its influence on changes in soil water content. If the soil beneath the slab is wetter than the soil outside the slab, then moisture moves from under the slab toward the slab perimeter. The distance over which this exchange in moisture content occurs is termed the *edge moisture variation distance*. The opposite happens if the soil beneath the slab is drier than the soil outside the slab. In this latter situation, water moves from the wetter soil outside the perimeter of the slab to the soil beneath the interior of the slab through a change in energy process associated with soil suction. Two principal types of slab distortion occur as a result of these two different types of soil moisture movement: center lift distortion and edge lift distortion.

- ***Center Lift Distortion***. Also called *s*, this distortion takes the shape of a plate bent with the edges down and the middle up. This type of distortion occurs as a result of the soil around the edge of the slab drying out and shrinking (Figure 1-17), or the soil beneath the center of the slab wetting up and heaving, or a combination of both. Sometimes the combination is produced by the soil beneath the slab's center wetting up and heaving while a tree or other vegetation is removing water from beneath the slab edge during a drought (Figure 1-18). This distortion mode is distinguished by floor slabs being distorted in a concave downward mode (e.g., Figure 1-19) and exterior cracks being wider at the top than at the bottom (e.g., Figure 1-20).

Figure 1-17. Gap caused by soil shrinkage.

Source: AccuTech Consultants, LLC, used with permission.

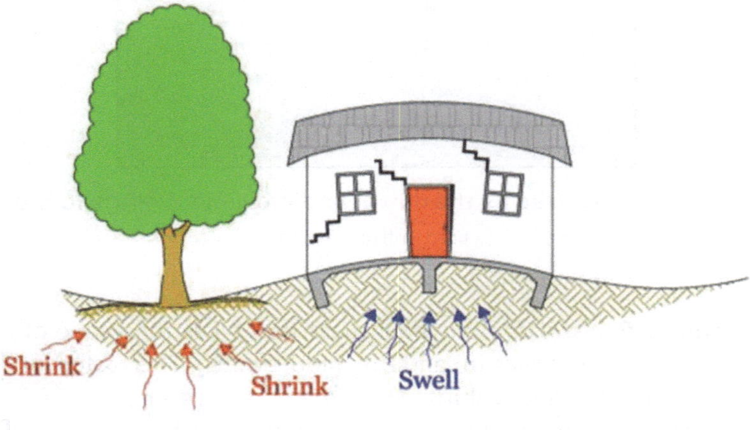

Figure 1-18. Center lift distortion that is typical, and the effect of trees that are planted too close to a building.

Source: AccuTech Consultants, LLC, used with permission.

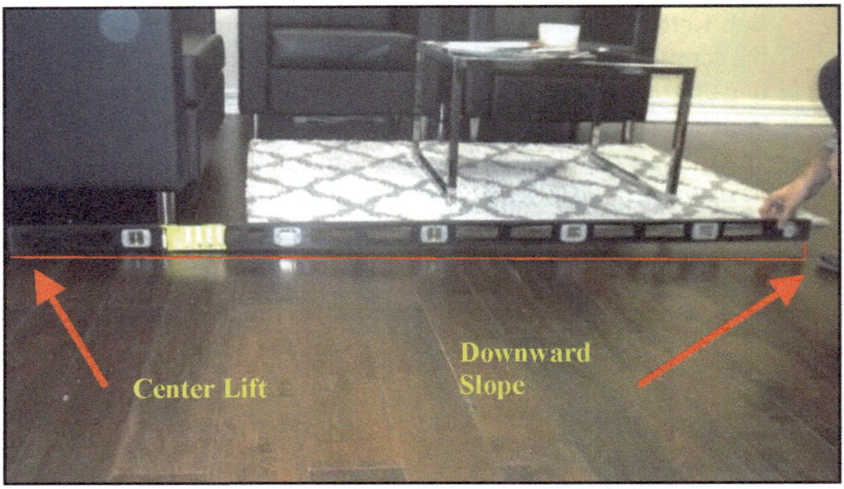

Figure 1-19. Floor slab distortion.
Source: AccuTech Consultants, LLC, used with permission.

Figure 1-20. Crack caused by the downward movement of the foundation.
Source: Marshall B. Addison, Ph.D., P.E., used with permission.

- ***Edge Lift Distortion.*** Edge lift distortion takes the shape of a plate bent with the edges up and the middle down. This distortion is also sometimes called dishing. This type of distortion occurs as a result of the soil around the edge of the slab wetting up and heaving (Figure 1-21).

Figure 1-21. Exaggerated depiction of edge lift or dishing.
Source: AccuTech Consultants, LLC, used with permission.

Although it could also be caused by the soil beneath the center of the slab drying out and shrinking, this situation seldom happens. This type of distortion mode is recognizable by exterior cracks being wider at the bottom than at the top (Figure 1-22).

Figure 1-22. Crack caused by the upward movement of the foundation corner.
Source: AccuTech Consultants, LLC, used with permission.

One Thing Leads to Another

Changes in soil moisture cause changes in soil volume, which in turn cause movement of the soil immediately below a home's foundation. The foundation then moves, which then can cause damage, such as cracking. Other types of distress include distortion of walls, floors, windows, and doors. This type of distress could take the form of gaps and separations or "sticking" or "jammed" windows and doors. Some distress is temporary or cyclic if the water content and soil volume change seasonally; gaps may close, and doors and windows may work normally at various times during the year. Recording measurements of movement will help in understanding the nature and severity of any problems you observe.

Interior Cracks. Many interior cracks are caused by the same sort of distortion that produces exterior cracks and are usually of three types: horizontal, vertical, and diagonal cracks.

Cracks that are essentially horizontal or vertical are typically following the taped joint between adjacent pieces of drywall or Sheetrock. Although these cracks may be indicative of foundation movement, they are often the result of the wood trim around doors and windows drying out. Even though the trim is painted, there are still enough exposed surfaces on each piece to permit the wood to dry out during the less humid periods of the year. The wood trim does not shrink evenly in the cross-grain and with-grain directions. The result often is the nailed trim shrinks and pulls the piece of drywall (to which it is attached) away from the adjacent piece of drywall, causing the drywall joint tape to split.

Diagonal cracks in Sheetrock walls, however, suggest that some foundation movement has occurred. To determine the type of distortion causing the diagonal crack, it is often helpful to look at the bottom of the baseboard or the top of the wall where the wall joins the ceiling. If the edges of the foundation are heaving (edge lift) and the roof is supported by trusses (Figure 1-23), the ceiling and the floor can often be observed to be separating from the wall (Figure 1-24).

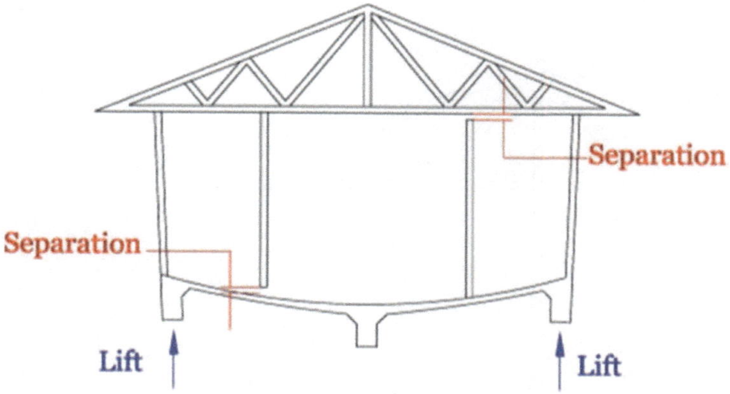

Figure 1-23. Foundation edge lift.

Source: AccuTech Consultants, LLC, used with permission.

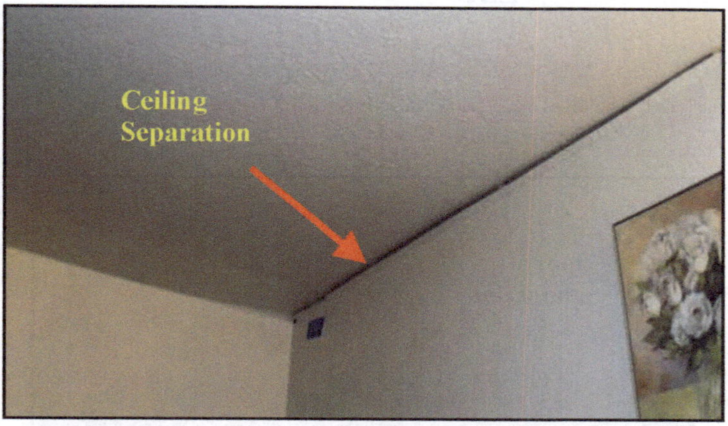

Figure 1-24. Ceiling wall joint separated at the interior partition because of slab edge lift.

Source: AccuTech Consultants, LLC, used with permission.

Another indication of the floor heaving can be obtained by observing a wall that is fastened to the roof truss separating from the floor (Figure 1-25) when the edges of the foundation are moving downward (center lift) (Figure 1-26).

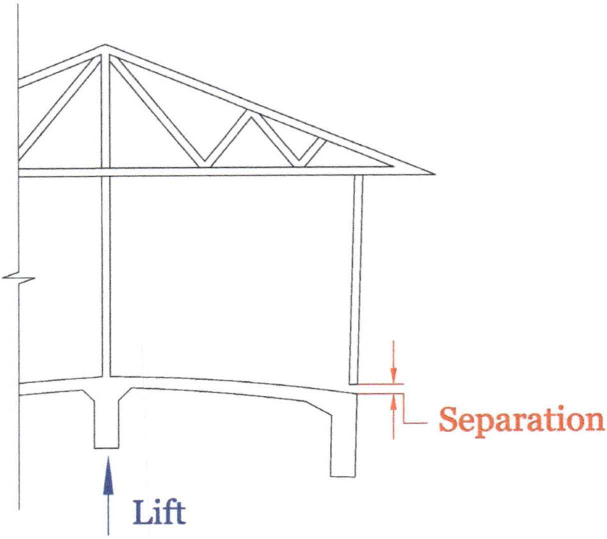

Figure 1-25. Foundation center lift.

Source: AccuTech Consultants, LLC, used with permission.

Figure 1-26. Floor/wall separation at perimeter wall owing to center lift.

Source: AccuTech Consultants, LLC, used with permission.

How Lot Drainage Affects the Performance of Your House

If a house is properly designed and constructed, and if the vegetation is selected and located with care, the house or building can still experience some severe damage as a result of poor or improper surface water drainage across the lot. With respect to reducing damage resulting from expansive soil movements, a major objective is to keep the soil from wetting up or drying out, that is, maintain a relatively constant soil water content.

©JSN

Drainage Tip

A home's lot drainage should direct surface water, usually rainfall runoff, away from the house or building and prevent water from collecting or standing adjacent to the structure.

REMEMBER

The lot drainage objective should be to direct surface rainfall runoff away from the house or building and to prevent water from collecting or standing adjacent to the structure (Figure 1-27).

Runoff from rainfall occurs principally from two sources: water running off the roof of the house and water flowing over the ground surface.

Roof Runoff. Water, whether in the form of rain or melting snow or ice, will either drop off the edge of the sloped roof or be caught in a gutter and discharged from a downspout. Flat-roofed buildings have other methods by which water is removed from the roof, but at some point or location, the water from the roof is discharged to the ground. Builders know that the ground surface adjacent to a house or building should be sloped away from the house or building—called a *positive slope*—so that the rainwater will not get a chance to soak the soil around the perimeter of the house or building. Many homeowners object to sloped lawns adjacent to their house.

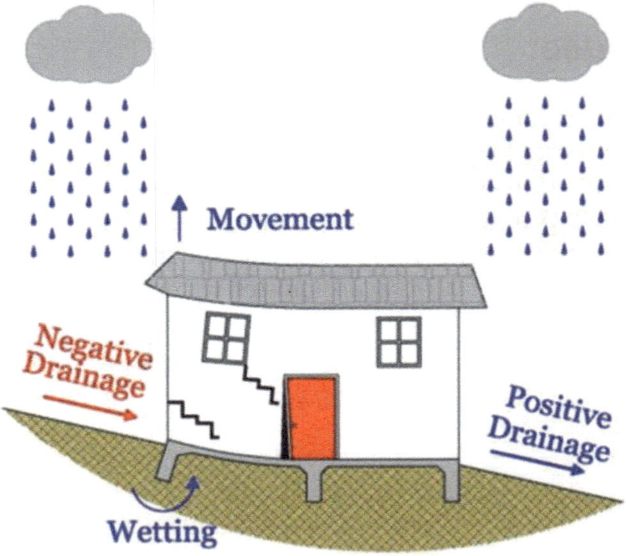

Figure 1-27. Poor drainage.

Source: AccuTech Consultants, LLC, used with permission.

Home Care Tip

If your home has gutters and downspouts, most experts recommend that drainage be directed at least 5 ft away from exterior walls. Downspouts are rarely extended this far when homes are originally built. Adding extensions to move water collected from the roof away from the foundation is a simple, inexpensive, and highly effective corrective action. Many homeowners with basements use this technique to help move water away from the foundations and keep the basements dry even without the added concerns associated with expansive soil conditions and movement.

However, the purpose of the slope is to ensure that water from the roof (or water from sprinkling or irrigation) does not pond adjacent to any portion of the foundation and cause the soil to heave around the perimeter of the house.

When gutters are used to catch the water as it runs off the roof, the downspouts that discharge the gutter water at ground level should not be permitted to discharge the water adjacent to the foundation (Figure 1-28).

Figure 1-28. Poorly placed downspout.
Source: AccuTech Consultants, LLC, used with permission.

Permitting downspouts to discharge their water adjacent to the foundation will result in the soil becoming very wet, and probably staying wet, and cause the soil to heave along the perimeter of the building. Most experts recommend discharging water from downspouts at least 5 ft (1.5 m) away from the foundation. A common complaint from homeowners who do not understand expansive soils is that having downspouts discharging so far away from the foundation creates an "ugly" site in their otherwise impeccable lawn, particularly if a splash block is placed under the exit of the downspout (to prevent a depression with standing water from occurring). There are also many complaints about the aggravation of having to mow or trim around the downspout (and splash block). However, experience tells us that the aggravation of temporarily moving the downspout and splash block for mowing is much preferred to the aggravation that is associated with a cracked wall in the house!

Surface Drainage. Every lot is sloped to some extent even though it may look "flat." The purpose of sloping the surface is to help runoff water move away from the house. Local experience and building codes vary from location to location, but most codes specify a minimum slope for paved surfaces of 1% (1 in. of "fall" in 100 horizontal in. [25 mm in 2.5 m]) for paved surfaces and 5% slope for grass surfaces (over the nearest 6 to 10 ft [1.8 to 3.0 m]). From experience, slopes "flatter" than these recommended minimums tend to let the water flow away too slowly, and the water either ponds in shallow depressions or soaks into the ground (Figure 1-29).

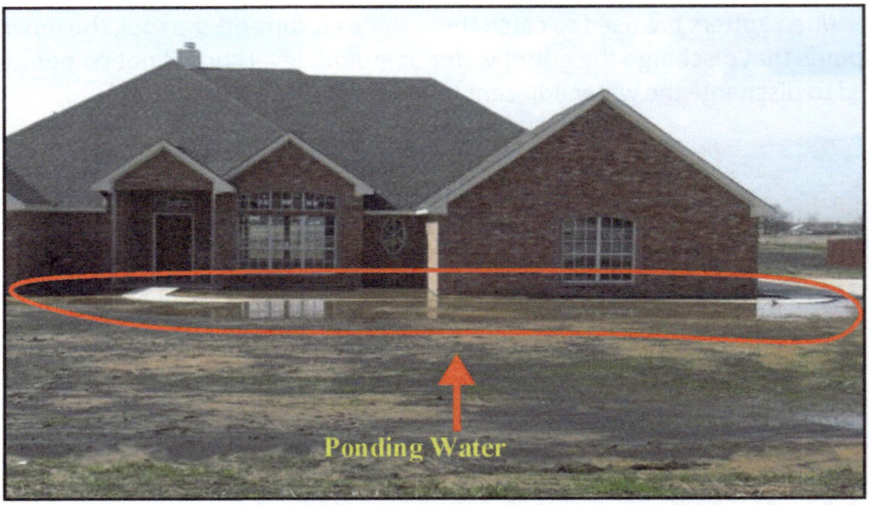

Figure 1-29. Poorly graded site that allows water to pond near the foundation.

Source: Marshall B. Addison, Ph.D., P.E., used with permission.

Sometimes the water from your neighbor's lot must flow across your lot to flow away to the street gutter, storm water channel, or a nearby stream. The builder or developer most likely knew that this would happen and graded your lot (sloped the surface) and probably each of your neighbor's lots (if the homes were all built by the same contractor) to accommodate this surface flow. Often, there is a tendency among homeowners to fill in the *swale* (a depression meant to serve as a minor water channel or to change the direction of the water flow) (Figure 1-30) because it otherwise interrupts a beautiful constant-slope lawn. Although this tendency is a natural one, you should never fill in a swale! The swale was deliberately built to ensure that water was directed around and/or away from your house. If the swale is altered in any fashion, then its purpose is defeated, and the result is likely to be wetter soil around your house and additional damage due to the heaving soil. If you have a swale across your lot and if it often "floods" (water spreads outside the boundaries of the swale), you should discuss the problem with your builder or developer. The swale may need to be widened, deepened, or the slope increased to facilitate the runoff water.

Houses with Basements or Below-Grade Floors. Houses with basements or floors below grade (lower than the outside ground surface) pose special drainage problems. The usual method of construction is to excavate a hole that is longer and wider than the finished dimensions of the basement. This overexcavation is to permit the builders to work on both sides of the wall.

On completion of the basement, that part of the excavation between the wall of the excavation and the outside basement wall is *backfilled* (usually filled with the soil that was removed from the excavation, or sometimes, an imported select soil). If the backfill is improperly accomplished or if surface water is permitted to infiltrate into the backfill, unwanted soil heaving, which can laterally deflect the basement wall, can occur (Figure 1-31). The simplest and most direct method of preventing surface water from infiltrating into the basement excavation backfill is to provide an adequately sloped positive surface so that the water will run away from the building and the backfill.

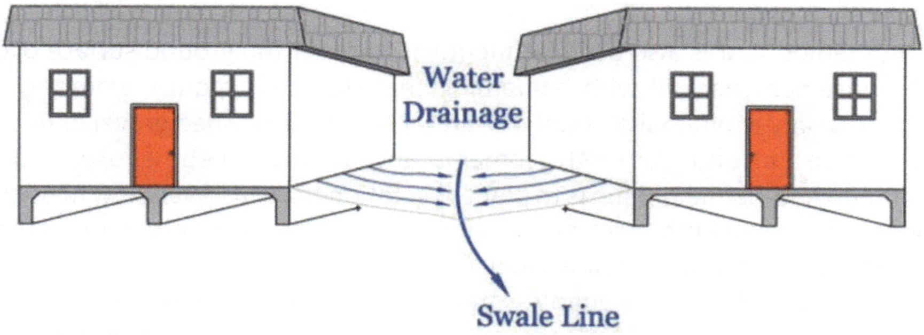

Figure 1-30. Drainage swale.

Source: AccuTech Consultants, LLC, used with permission.

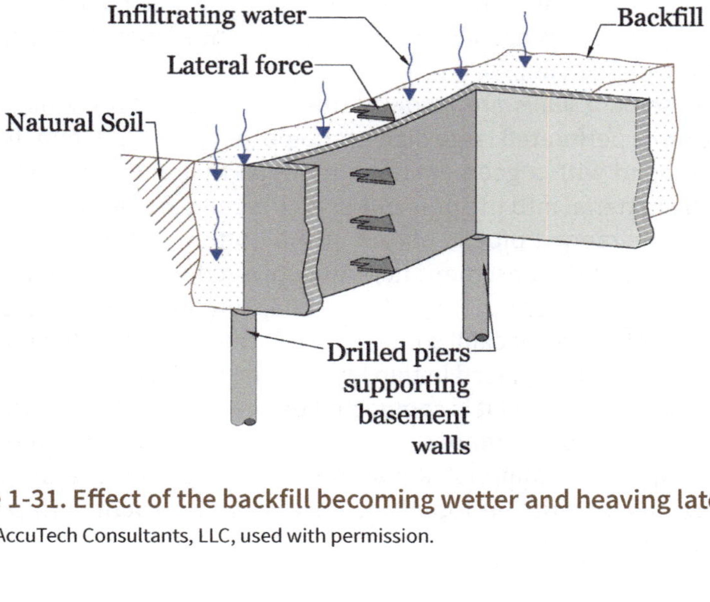

Figure 1-31. Effect of the backfill becoming wetter and heaving laterally.

Source: AccuTech Consultants, LLC, used with permission.

Another method that is effective when the width of the backfill is relatively narrow is to cover the backfill with something impervious. Sometimes a concrete sidewalk on the surface of the ground and poured against the outside wall of the house or building can be used to accomplish this objective. Often, a thick layer (typically 2 ft [0.6 m] thick) of compacted clay that is not allowed to crack will provide the proper protection against infiltration. A common practice is to put a geomembrane fabric beneath the topsoil layer placed over the compacted clay layer to intercept and carry away any surface water infiltrating through the topsoil layer; the fabric should extend at least 3 ft (1 m) beyond the edge of the backfilled basement excavation and be sloped away from the house.

A method that is also used frequently is to cover the ground surface over the basement backfill with something that does not require watering or requires only a minimal amount of water. Decorative crushed gravel or decorative tree bark underlain with geomembrane fabric is sometimes used in this manner. Another technique is to plant vegetation that has a very low need for water, thus reducing or even eliminating the need for artificial watering in the area between the house foundation and the lawn.

Another method of draining water away from a foundation is to use a peripheral drainpipe. This method is useful because it allows infiltrating water to seep into the perforated drainage pipe and be subsequently drained away from the foundation. Although this method could be used with a sump and/or pump, it is more frequently used for shallow foundations when the piping system can be drained by gravity flow. The danger in using this method is that a poorly installed system can result in water ponding in the pipes and draining back out into the soil, feeding excessive soil swelling. Thus, it is usually best to let experienced installers with references do the job for you.

When basement walls are involved, the drainage system is more complex. Slotted or perforated large diameter pipe (e.g., 4 in. [100 mm] in diameter) is wrapped with a geomembrane material that allows water to pass through the material into the pipe but prevents soil particles from entering the pipe. The wrapped pipe is placed at the bottom of the backfill space below the level of the basement floor and covered with several inches of crushed stone to prevent the pipe from collapsing from the weight of the soil above it before the backfill soil is placed in the excavation (Figure 1-32). The backfill soil is then placed in thin layers (*lifts*) (typically 4 to 6 in. [100 to 150 mm] thick), and each lift is compacted before the next lift is placed. The compaction operation continues until the excavation is filled to the height where an impervious material is placed over the backfill. The purpose of the impervious material on top of the backfill is to prevent surface water from infiltrating the backfill.

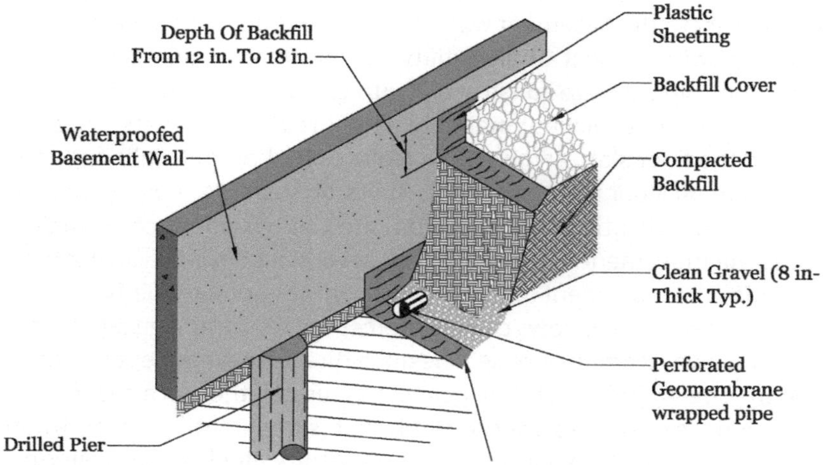

Figure 1-32. Perforated pipe drainage system.

Source: AccuTech Consultants, LLC, used with permission.

There are four principal reasons for not wanting water to enter the backfill: (1) Basement walls are not built to withstand the lateral forces exerted by water on the backside of the wall (called *hydrostatic pressure*). (2) Water that enters the backfill might find its way into the basement. (3) Infiltrating water will feed the expansive soil beneath the basement wall and under the basement floor and cause the basement wall to heave and the basement floor to break. The floor will then heave if separated from the basement wall. If the basement floor heaves, the movement typically results in the basement stairs being pushed upward, plumbing lines being broken or pushed upward, and if the interior basement walls rest on the basement floor and connect to the structural members supporting the ground floor, the heaving basement wall will cause the upper floors of the structure to become distorted, too. (4) To avoid getting the backfill soil wet and producing a lateral swell pressure against the basement walls. Gravity drains are the most reliable systems, but often sump and pump systems must also be used if sufficient change in elevation cannot be found to provide the gravity drainage. If the sump and pump system is used, the discharge pipe from the pump must be directed away from the building and the pipe must discharge into an appropriate storm water collector or it must discharge onto the ground (if codes permit) far enough away from the building that the water cannot run back into the backfill!

Owners of houses with basements in expansive soil should expect the basement floor to move up (or down), particularly during the first few weeks after construction is completed. The reason for the movement is that the soil beneath the floor slab is coming into a soil moisture equilibrium condition during the period following completion of construction. Usually the engineer or architect who designed the building also designed the floor slab to move with respect to the basement walls and any columns that penetrate the floor as well as with respect to any utility penetrations (water, gas, sewer). The utility lines typically are constructed with collapsible sleeves that permit the floor to move vertically but maintain a tight relationship between the utility line and the floor slab. Architectural features that permit walls supported on the basement floor slab, as well as stairs or steps, to move vertically with the slab have also been incorporated into the basement construction and, unless the movement is excessive (i.e., several inches), the owner will likely not notice the movement if the construction was done properly.

Sometimes surface flows can be intercepted by a drainage ditch excavated at a point away from the house. The ditch directs the intercepted water away from the site. A similar solution can also be used if there is a shallow stratum of pervious material (e.g., silt or sand) that is transmitting water toward the location of the house. Again, by intercepting the water from the water-bearing stratum by either an interceptor ditch or a buried perforated pipe, the water that would otherwise produce soil movement beneath the house is captured and directed away from the site.

Investment Tip

The best solution, if you are building a new home in an area with expansive soils or with unusual site characteristics such as rock and boulders, sloping ground, or water nearby, is to engage a qualified professional engineer to help evaluate your building site and design an appropriate, safe, functional, and economical foundation for the house or building.

Building on a Sloping Site

In this discussion, a "flat" site is assumed to be one that only appears to be level but in reality has enough slope that water will drain away from the house or building. With this definition, many of the previous sections have already discussed the drainage aspects regarding home sites. However, building on a hillside or otherwise steeply sloping site creates opportunities for different kinds of structural performance.

The first, and perhaps the most important consideration, in building on a slope or a hillside comprised of expansive soil, is surface drainage. It is very important that the surface drainage not change the subsurface soil water conditions on your site. This means that surface runoff from your lot not only must drain the water quickly and effectively from your lot, but the water from your lot (or the cumulative water from several upslope lots) must not impact the subsurface soil water conditions on your neighbor's lots. In addition, if there is subsurface drainage occurring across your lot, any construction that you do must not impound or block the flow of the water beneath the surface. A typical solution is to use buried slotted or perforated pipes.

If the sloping site has had minimal grading done to it and the house is essentially a "split-level" structure, the house or building is susceptible to the same movements from shrinking and heaving expansive soils that a house or building constructed on a level site experiences—with one exception. As expansive soils on a slope swell, the individual clay particles are constrained from swelling laterally by the adjacent clay particles (as described previously in "Heave"), which are also trying to swell laterally. Thus, the soil is compelled to swell upward, as explained previously. However, the ground surface on a slope is not horizontal, but sloped, and the "upward" swelling is actually not quite vertical, but at an angle to the vertical and perpendicular to the sloped surface. When the shallow soil begins to dry out and shrink, gravity induces the shrinking clay particles to move downward vertically, not perpendicular to the sloping surface. As the swelling and shrinking cycle is repeated season after season, the effect of the nonvertical heaving and the vertical shrinking tends to cause the shallow clay particles to migrate downhill in a sort of saw-tooth fashion as depicted in Figure 1-33. When all the clay particles involved in the sawtooth shrinking and swelling phenomenon are responding in this fashion, the gross result is that the whole slope is slowly but surely moving downhill, carrying anything supported in or on its mass with it!

One way to prevent slow downhill movement is to construct a level surface (a *pad*) on the hillside. This is typically done in one of two ways. One way is to transport in nonexpansive soil and build up a level building surface. The result is a wedge-shaped mass of soil (a *fill*), thicker on the downhill side and thinner on the uphill side. The original sloping surface is usually *benched out* (the slope is excavated in a stair step fashion) with each level sloped slightly downward and backward into the hillside. The bottom bench is excavated about 2 ft (0.7 m) below the natural or finished ground surface to provide a "keyway" to further resist downhill movement. The good thing about this solution is that the nonexpansive soil eliminates the shrink–heave problem that results from your foundation bearing on an expansive soil. However, the expansive soil is still present below the new, nonexpansive fill material. Usually nonexpansive soil is more permeable than expansive soil and will tend to collect water under the house unless care is used to prevent and drain water from the nonexpansive fill layer. If the underlying expansive soil becomes exposed to changes in soil water content and begins to experience cyclic episodes of shrinking and swelling, the expansive soil will begin to move downhill in that now-familiar sawtooth fashion, and it will carry the new, nonexpansive soil and your house or building with it, just as before.

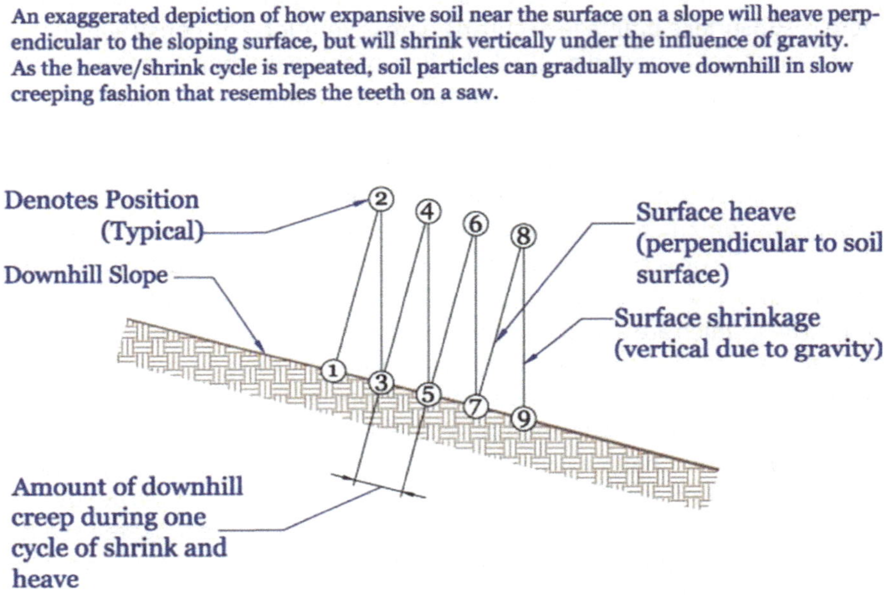

Figure 1-33. Sawtooth shrinking and swelling.

Source: AccuTech Consultants, LLC, used with permission.

Another way to avoid the unpleasant experience of slowly moving downhill while sleeping in your bedroom is to construct a level surface on the hillside by excavating part of the hillside and using the excavated material downslope of the excavation to build up a fill. The combination of excavation and fill material creates a level building pad surface known as a *cut-and-fill site*. This sort of solution can create some different problems for the house or building. One problem created by this method is that the expansive soil near the surface is removed from the uphill part of the site during the "cutting" operation. However, depending on the stratigraphy, not all of the expansive soil may be removed; new expansive soil may be exposed.

A second problem created by this cut-and-fill solution is that the newly exposed expansive soil will not only shrink vertically but will also heave vertically because of the now level ground surface. The removed expansive soil that comprises the downslope fill will also shrink and heave vertically, but if the expansive soil underlying the new fill experiences soil water content changes and begins to shrink and heave, the fill material can begin to move downhill along with the underlying soil, as described above. The principal problem with this situation is that the part of the house or building foundation that is embedded in the cut section of the building site will tend to "stay put" and only move up and down as the supporting soil shrinks and swells, but that part of the foundation embedded in the fill section will be forced to follow its supporting fill soil downhill (Figure 1-34). The ultimate result of this cyclical, long-term soil movement is that the house or building will tend to experience damaging strains.

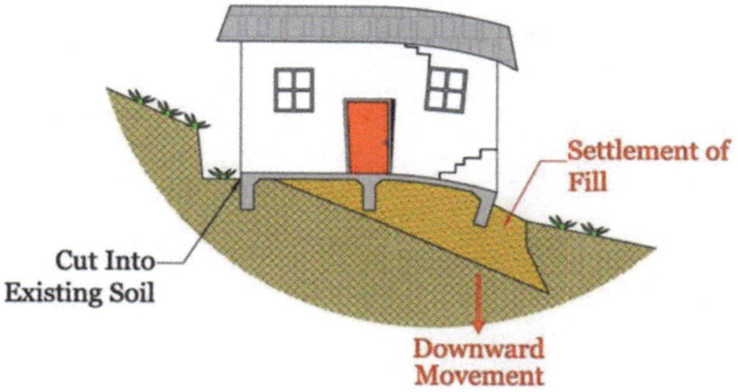

Figure 1-34. Settlement of fill resulting in cracks in the exterior wall finish.
Source: AccuTech Consultants, LLC, used with permission.

A third solution to constructing a building or house on a hillside is to use deeper foundations which penetrate through the upper few feet of soil and transmit the structural loads to a more stable soil at depth.

REMEMBER

The best solution, if you are faced with building on a sloping site comprised of expansive soil materials, is to engage a consulting geotechnical engineer to make building site and foundation recommendations.

Part 2 Frequently Asked Questions and Their Answers

How to Use "Part 2"

Part 2 of this book is organized as questions and answers by topic. This section provides additional and supplemental information, suggestions, and technical content to complement the overview provided in Part 1. Additional information can be found using a number of online resources by searching for information on "expansive soils," "swelling soils," "shrink-swell soils" or similar terms. For in-depth details, technical papers on particular aspects of these soils and their behavior may be available. A reference librarian at your local library will be happy to assist you. Many libraries subscribe to periodicals and journals with detailed technical research and analysis content.

This section will address some common questions that homeowners and building owners often ask. The questions and answers are grouped in generalized headings.

About Expansive Soil

What is an expansive soil?

An expansive soil is a clay soil that has the ability to change in volume when the water content of the soil changes (Figure 2-1). The soil will shrink when the soil water content is reduced, and the soil will swell when the water content increases. A soil is commonly considered to have expansive tendencies when its *plasticity index* (PI) is greater than 20. If the PI is greater than 20 but less than 40, the soil is considered to have moderate expansive properties. The soil is considered to be highly expansive if the PI is between 40 and 60. Soils with PIs greater than 60 are considered to be very expansive. Another method often used to classify the expansive potential of soil is the *expansion index* (EI). A soil with an EI of 50 or less is considered to have a low expansion potential. An EI of 91 or greater indicates a soil with high or very high expansion potential.

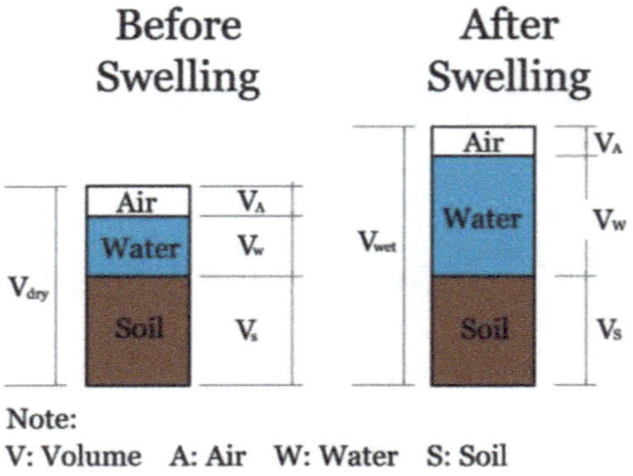

Figure 2-1. Soil shrink-swell properties: the volume of water changes with additions or depletion of soil moisture, which affects the total soil volume.
Source: AccuTech Consultants, LLC, used with permission.

What is the plasticity index?

The plasticity index (PI) is an engineering term that measures the difference between a soil's *liquid limit* and *plastic limit* (Figure 2-2). The liquid and plastic limits are soil water contents and are defined by special standard geotechnical engineering laboratory tests. The PI defines a range of soil water content over which the soil remains in a plastic or moldable state. If the soil's moisture content is less than the plastic limit, the soil is considered to be in a dry state and will exhibit cracking. If the soil's moisture content is greater than the liquid limit, the soil is considered to be very wet and likely will have low strength properties and will be very compressible under load. The liquid limit, plastic limit, and plasticity index do not measure any particular soil property, but their respective values have special meaning to geotechnical engineers.

What is the expansion index?

The expansion index (EI) is used to measure a basic index property of a soil. It is comparable in function to the liquid limit, plastic limit, and plasticity index

of soils. The EI is determined from a special laboratory test that is performed in a specified standard manner. It has a special meaning to a geotechnical engineer.

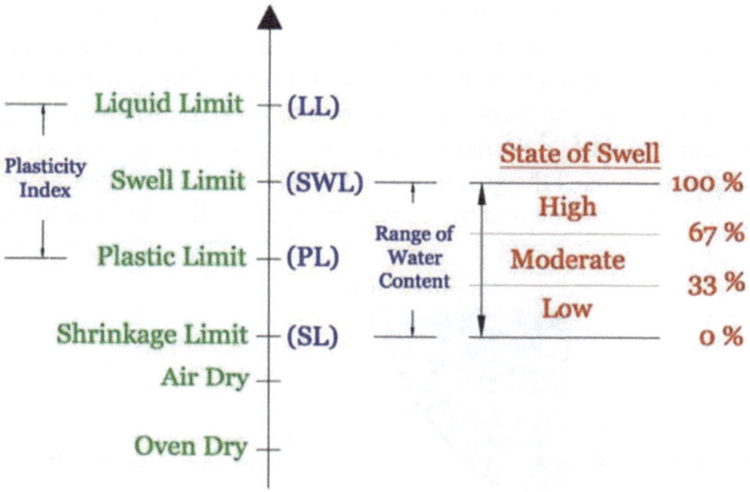

Figure 2-2. Soil shrink-swell properties.
Source: AccuTech Consultants, LLC, used with permission.

Why are expansive soils "expansive"?

The clay particles that compose expansive soils are different, depending on the type of clay mineral. However, the clay mineral that has the greatest ability to expand and contract is a type of clay mineral called *smectite*. Smectite soils, which include such specific clay minerals as *montmorillonite* and *bentonite,* have a surface electrical charge that attracts water molecules and free cations, which have attracted free water molecules themselves (Figure 1-2). The individual smectite clay particles are weakly bonded to each other, and water molecules have the ability to readily break this bond and insert themselves between the previously bonded clay particles. If the soil becomes wetter, more water molecules are attracted to the clay particles and the water molecules push the adjacent clay particles farther and farther apart, resulting in an increase in volume of the soil mass, or *swelling* (expansion, heave). The opposite happens when the soil begins to dry out and gives up moisture. The loss of water molecules permits the clay particles to move closer together and the result is a reduction in volume of the soil mass, or *shrinkage*.

What is an active zone?

Measured from the surface downward, there is a depth over which expansive soils experience a change in soil moisture conditions as the climate (or seasons) change, which results in the soil actively shrinking or heaving (Figure 2-3). This *active zone* could be only 2 or 3 ft (0.6 or 0.9 m) deep or could be 15 ft (4.6 m) or deeper, depending on the climate. The active zone thickness is influenced by the location of the groundwater table. If the groundwater table is shallow (i.e., near the surface), it will tend to keep the soil near the surface wetter than it would be if only the climate were influencing the surface soil moisture conditions. If the groundwater table is deep, it will have a negligible effect on the active zone and the climate will govern the depth of the zone.

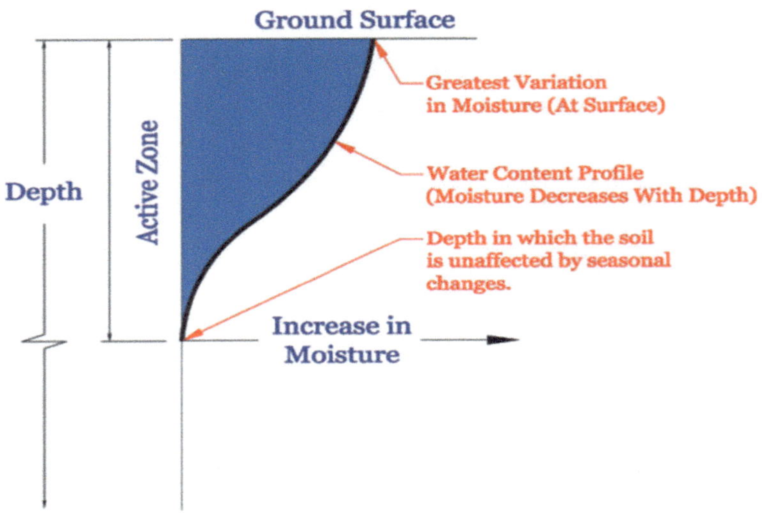

Figure 2-3. Soil active zone.

Source: AccuTech Consultants, LLC, used with permission.

What is the difference between groundwater table and perched groundwater table?

The *groundwater table* is a name given to the true groundwater table or the free water surface of the water in the ground. It is the level or elevation to which the water will rise in an open hole. A *perched groundwater table* is not the true groundwater level. Typically, a perched groundwater table is free water encountered in the ground that is created when an impervious layer of soil or rock lies at an elevation above the true groundwater table and prevents water from percolating from the surface down to the true groundwater table. If the perched groundwater table is near the surface, it can artificially control the depth of the active zone in the vicinity of the perched groundwater table.

About Expansive Soil Behavior

What can cause the soil to shrink or swell?

We know that the expansive soil swells when it gets wetter and that it shrinks when it dries out, but there are a number of reasons for or ways that a soil can get wetter or drier.

In climates that experience seasons that have a definite wet period and an equally defined dry period, the climate affects the soil water content. Obviously, the soil is wetter during and immediately following the "rainy" season, whereas the soil will usually be drier during and immediately following the dry season. Sometimes a long-term dry spell, or *drought*, will impact a location. During drought periods, the soil will become drier than normal and will dry out to a deeper depth than normal (causing a deeper active zone). A drought will induce a greater distortion in a foundation and the superstructure that it is supporting, resulting in increased and/or greater damage than usual.

Plants, trees, shrubs, and others will attempt to supply their water needs by extracting water from the soil through their roots. During dry seasons, these plants (particularly trees) can virtually desiccate a soil as they attempt to remain alive. Sometimes, however, the drought is so severe that the plants either exhaust all of the available water from the soil or reach the limit of their ability to remove water from the soil. When either of these situations occurs, the plants wilt and become dormant. If this period of dormancy lasts too long, some plants will die; others will revive when the drought is broken and water is once again available. During the drought periods, when the plants are removing all the available water that they can to stay alive, the soil will shrink as it dries out. If the trees, shrubs, or bushes are near the house and its foundation, the trees (shrubs or bushes) can cause the house slab perimeter to bend down. When the drought is over and water is once again available in the soil, the soil will expand as it gains water and becomes wetter.

The soil also gives up water to the atmosphere in the form of evaporation. This phenomenon provides an additional impact during periods of drought. Plant transpiration, as described in the preceding paragraph, can actually remove more water from the soil than can evaporation during periods of drought.

How quickly does shrinking and swelling occur?

The shrinking and swelling phenomenon will occur almost instantaneously at the point of contact between the free water and the clay particles. However, most clay soils have a very low soil permeability or conductivity. If the source of the water increase remains or is available frequently, and there is a sufficient amount of water available, water will gradually flow deeper into the soil mass. Increasing depth of wetting occurs either under the influence of gravity, as a result of a difference in soil suction, or because of a difference in soil temperature.

What is soil suction?

Soil suction is an energy term that, in nontechnical terms, measures a soil's affinity for water. The greater the soil suction, the drier the soil, and vice versa. A saturated soil will have almost no soil suction. Geotechnical engineers use soil suction to evaluate soil moisture conditions and swell potential for foundation design.

If the source of wetting or drying is removed or terminated, will the heaving or shrinking stop?

The swelling or shrinking can occur over a long time: weeks, months, or even years. The swelling or shrinking will continue until the soil reaches some equilibrium and the soil has either imbibed or given up all the water that the soil moisture conditions permit.

Engineer's Tip

Even if a cause of wetting or drying is removed, heaving or shrinking can continue for a long time afterward—weeks or even months—although the magnitude (measured movement) of shrink or heave is expected to reduce with time. The type of soil, amount of water, site geometry, and other factors will determine how the soil responds to changes such as redirecting surface water, building a home addition, or removing or changing vegetation.

The swelling and shrinking can also be cyclic in nature, particularly if the reason for the swelling or shrinking is related to changes in climate or due to vegetation.

How much will the soil heave or shrink?

The magnitude of shrink or swell is dependent on a number of factors. The following factors most frequently impact the amount of soil volume change:

- Mineralogical composition of the clay soil,
- Amount of clay in the soil,
- Soil structure and fabric,
- Initial soil water content or the initial soil suction,
- Soil and/or water chemistry,
- Initial density of the soil,
- Thickness of the expansive soil stratum,
- Location of the expansive soil stratum (more shrink or heave if the stratum is near the ground surface),
- Thickness of the active zone,
- Soil permeability or conductivity, and
- Site climate.

Amounts of shrink or heave have been measured and reported that range from negligible amounts to more than 24 in. (approximately 610 mm) in an extreme case.

What is the difference between total soil movement and differential soil movement?

Often, the magnitude of the total shrink or heave does not create as much of a problem for a structure as does the differential shrink or heave. Some structures can experience several inches of *total soil movement* without experiencing much damage if the heave occurs uniformly, but the same structure might experience considerable damage from only a fraction of an inch of *differential movement.* Total soil movement is a measure of the gross or maximum amount of shrink or swell that the soil experiences as its moisture conditions change; it usually refers to vertical soil shrink or heave.

Differential movement occurs when the heave (or shrinkage) is different between two points in the structure (Figure 2-4); differential soil movement usually refers to vertical soil movement, too. Differential movement creates strains and stresses which, if sufficiently large, result in cracks and other damages to the structure and the building's utility systems. If the differential soil movement is the result of climatic changes, then most of the differential movement will occur around the perimeter of the structure if the foundation is a slab-on-grade. The observed long-term distortion mode is center lift or doming, particularly for basement structures, and the differential movement can approach the total heave in some instances.

Figure 2-4. Downward movement due to shrinkage.
Source: AccuTech Consultants, LLC, used with permission.

What will be the pattern of volume change?

If the soil around the edge of the building dries out, as it would during a dry period or drought, then the building foundation would be distorted into a shape in which the edges are bent down with respect to the center of the slab. This same distortion mode will also occur if vegetation, especially trees, have their roots penetrating under the edge of the building, removing water during dry periods. This type of distortion is called center lift or doming. If the soil around the edge of the building becomes wetter, as it might during the rainy season, then the building foundation is distorted into a shape in which the edges are bent up with respect to the center of the slab. This type of distortion mode is called edge lift or dishing (Figure 2-5).

The most common distortion that occurs is not symmetrical. In fact, it is not unusual to find a part of the building to be heaved up while another part of the building is distorted downward. These different types of distortion might be due to excessive water entering the soil at one location while a large plant, such as a tree, is removing water from the soil at another location; changes in soil characteristics over just a few horizontal feet are not uncommon, and the differential distortion may even be the result of different types of soils or changes in soil stratigraphy between the two points. The thing to remember regarding the pattern of soil movement is that uniform soil movement is not the typical situation.

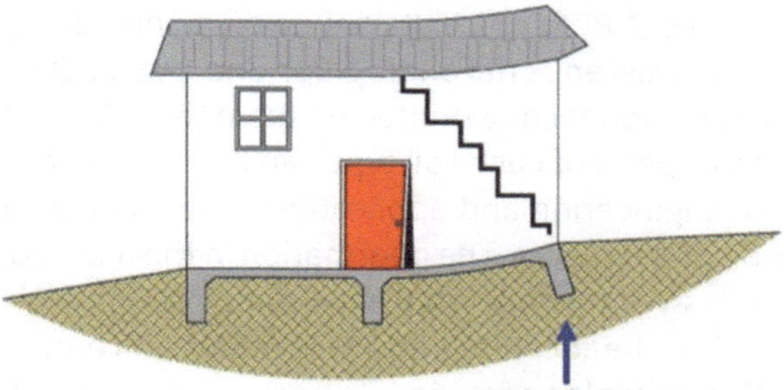

Figure 2-5. Upward movement due to swelling.
Source: AccuTech Consultants, LLC, used with permission.

Can enough downward force be placed on expansive soil to resist upward movement?

Swelling soils have been measured to produce swelling pressures of several tons per square foot. Residential structures and other 1- or 2-story buildings seldom weigh enough to resist the upward movement of the soil swelling beneath them. Even if the structural loads could somehow be concentrated, the downward-acting structural loads are seldom sufficiently large that they could counteract the upward acting forces of the swelling soils.

About the Cost of Building on Expansive Soils

Is it more costly to build on an expansive soil site?

Typically, it costs more to build a house on an expansive soil site than it would cost if the site did not contain expansive soils. One of the principal reasons for the higher costs is that the foundation must be more substantial to accommodate the soil movements and the degree of the soil's potential for volume change. Another reason for higher costs is that if the soil has a large potential for volume change (i.e., several inches of heave are expected), a completely different type of foundation may have to be used to support the structure and prevent it from being damaged when the soil experiences shrink and swell changes in volume.

About Expansive Soil Locations

How can I find out if my house is located on expansive soil?

A quick method of determining if your house or building site is on expansive soil is to obtain a county soil report from the local USDA Natural Resources Conservation Service (NRCS) office. Its online Web Soil Survey allows you to insert an address to view an aerial photograph and map of the area. The maps will have a number or letter code printed on them that identifies a soil type. A short, written description of each soil type can be found at that website, and several tables of engineering and applications information can be found in the tables section of the report. The information in the text and the tables will identify the shrink-swell tendencies of each soil type.

Calling or visiting the local building inspection office or the city/county engineering office may also provide information about the type of soil on your house site.

The most accurate and definitive way to determine whether expansive soils are present on your house or building site is to hire (retain) a geotechnical engineer to investigate your site.

Engineer's Tip

Although it costs more to build on a site with expansive soils compared to a site without expansive soils, if that is the location where you want to build, those costs are necessary to help prevent problematic deformation and damage, which costs much more to fix than the added initial cost of the foundation. It also costs more to build on hillsides, in areas with deep frost penetration, and on soft ground; in these cases, too, a properly designed foundation is an important investment in the quality and longevity of the rest of the house or building.

Where can I get more information about expansive soils?

Civil engineering departments at universities usually have textbooks that are informative and faculty members who are knowledgeable about expansive soils, particularly if there are expansive soils present in the general area of the university.

Online searches with keywords found in this book will provide a wealth of additional information. University libraries are more likely to have access to technical journals and technical conference proceedings, which often have articles published about research involving expansive soils.

The US Department of Housing and Urban Development (HUD) in Washington, DC, also has printed information on expansive soils available.

About Hiring an Engineer

Should I hire a professional engineer if my house or building will be constructed over expansive soil?

Most reputable, experienced home developers know to involve professional engineers in designing the foundations for houses to be built on expansive soil sites. However, if your new house is going to be "custom built," you should make sure that a registered professional engineer who specializes in geotechnical engineering is *retained* (engineers, like other professionals, are "retained" rather than hired) to investigate the subsurface soil stratigraphy and the soil properties on the site and to provide foundation recommendations.

Homeowner Tip

Home developers often use several professionals (architects, surveyors, and engineers) when they design new housing communities or apartment and condominium complexes. Professional engineers can help homeowners building from plans or doing custom work when there are concerns or problems associated with a home's foundation or structure. If problems arise, engineers can help locate the cause and suggest corrective actions to solve the problem.

What type of professional engineer should I retain?

Typically, there are two types of professional engineers involved in the design of your house. The geotechnical engineer will evaluate the site stratigraphy and the soil properties obtained from the laboratory tests and, based on the test results and professional experience, provide foundation recommendations to a registered professional structural or architectural engineer. Some geotechnical engineers also do foundation structural design, but they more typically are engaged to provide recommendations to a structural or architectural engineer. The structural or architectural engineer will use the geotechnical engineer's recommendations to design the appropriate foundation for your new structure.

How much does it cost to employ a professional engineer?

The cost of a soil investigation, laboratory testing, and a written report will vary, depending on the geographical region in which the investigation is being accomplished. It will also depend on how far the site is from the geotechnical engineer's laboratory; the engineer will charge a mobilization cost to move the subsurface exploratory equipment to and from the site. The site investigation cost also depends on the number of and the depth of the exploratory borings; this cost is usually on a "per foot" of boring basis, but hourly charges are also common. Each laboratory test performed will have a specific cost. The geotechnical engineer will have a schedule of fees for services and the engineer can provide you with an approximate cost before any work is done. The engineer will also charge a fee for evaluating the results of the borings and laboratory tests and the recommendation that will be made to the structural or architectural engineer in the geotechnical engineer's written report. Simple sites will require fewer borings and fewer laboratory tests, whereas more complex sites will require more exploration and more tests.

The structural or architectural engineer typically charges by the hour. Their time to design the foundation structure will depend on the type of foundation, the type of reinforcement that will be used, and the complexity of the house or building floor plan. It is difficult to estimate in advance what the structural or architectural engineer's fee will be but plan for the geotechnical and structural/architectural engineers' fees to be approximately equal.

What should I do if I need additional information?

If additional information is required on any of the topics contained in this book, we recommend that you reach out to your local branch of the ASCE for further direction. For example, consider the following:

- *Guidelines for the Evaluation and Repair of Residential Foundations* (ASCE 2008),
- *Living with Shrink/Swell Soils: The Homeowners' Guide to Maintenance of Shallow Foundations* (ASCE 2016),
- *Home Construction on Shrinking and Swelling Soils* (Holtz and Hart 1978),

- *Home Landscaping and Maintenance on Swelling Soil* (Jochim 1987), and
- *Guide for Performance Evaluation of Slab-on-Ground Foundations* (PTI 2018).

Resources Tip

The American Society of Civil Engineers has local and regional chapters across the United States (https://www.asce.org). Geotechnical engineers, geologists, or other professionals are available to assist with information on your regional geology and local design practice. Homeowner resources (books, pamphlets, web pages) are also available related to drainage, low water consumption vegetation, and architectural landscaping.

About Types of Foundations

What types of foundations are used in expansive soils?

Foundation types can be categorized as "deep" foundations and "shallow" foundations. Pier foundations are usually considered to be *deep* foundations. Slab-on-grade (often called slab-on-ground; sometimes called a "waffle" slab or a "ribbed" slab, or less desirably, a "mat") foundations and spread footings are usually considered to be *shallow* foundations. The majority of foundations used on expansive soil sites are one of the following:

- Slab-on-grade,
- Slab-on-grade with piers,
- Post-and-beam,
- Basement with wall footings and slab floor,
- Basement with wall piers and slab floor, and
- Structurally suspended floor slab on piers.

Slab-on-grade foundations are usually constructed with either mild steel reinforcement (steel "rebar"), or with steel cables as the reinforcement. Mild steel-reinforced slabs are constructed with the common steel rebars placed inside the concrete forms and then the concrete is poured over the bars. Post-tensioned slabs are constructed with steel cables covered with a plastic sheathing instead of using steel bars. The cables extend through the sides of the forms. After the concrete is poured over the cables and has hardened, the steel cables are stretched or tensioned and then fixed so that they cannot shrink back to their original length. Post-tensioned slabs usually have larger cracks in them initially than either welded wire fabric or steel rebar reinforced slabs, but upon tensioning the cables, the concrete's drying shrinkage cracks in the slab are squeezed together and the slab's concrete is placed in compression.

Which type of foundation is best?

Both mild steel and post-tensioned types of slabs can work equally well, but each type has a different technical objective. A structural engineer, using the recommendations of a geotechnical engineer, should have the principal responsibility for choosing the foundation system for your house.

About Preventive Measures

Is there a best time of year to construct the foundation?

Yes, there is a best time to build. This best time is when the soil is neither at its wettest nor at its driest condition, that is, when the soil is near "equilibrium." However, it is usually impractical to wait for this optimum time to start building. Thus, it is important that your foundation be designed and built so that it will perform adequately under all of the normally expected conditions that the house will experience during its usable lifetime.

What type of backfill should be used under and around my house or building?

If the foundation is a shallow foundation, one solution that has been used successfully is to remove the expansive soil down to a specified depth and replace it with a nonexpansive soil. However, this is usually an expensive way to solve the problem. Sometimes the terrain lends itself to constructing an elevating pad of foundation soil beneath the foundation without causing the foundation cost to be exceptionally great. When soil is used that is different from what already exists at the building site, that soil must be nonexpansive.

A *nonexpansive* material should have a liquid limit less than 35 and a PI between 4 and 15. Sandy or coarse silty soils meet this criteria, but because of their relatively high permeability or conductivity, if water should ever be introduced into this type of soil, it could travel throughout the soil mass, wet the underlying expansive soil, and cause the heaving that the nonexpansive soil was meant to prevent. Therefore, if select fill is placed above expansive soil as a building pad, specialized recommendations should be made by your geotechnical engineer to address this situation.

If the house has a basement, the backfill soil ideally should be a nonexpansive clayey soil, not a sandy soil. Steps should be taken to prevent water from entering the backfill regardless of whether the soil is expansive or nonexpansive. If the backfill is expansive, then unwanted lateral swelling pressures will be imposed on the basement wall. If the backfill soil is nonexpansive and if a considerable amount of water collects in the backfill, the water will impose a against the wall. Water collecting behind the wall can cause damage because basement walls are seldom designed for hydrostatic pressure unless the basement extends below the groundwater table.

Will ponding the foundation soil before construction eliminate the swell problem?

The principle behind the idea of "ponding" is to flood the foundation site and keep it flooded for several weeks so that the water percolates down into the expansive soil and causes it to "pre-heave." Although it is a good idea in principle, it has never been successfully applied. Ponding also presents a different problem: If the ponding operation is applied for several weeks, the result will be very wet soil near the surface. The foundation will have to bear at a depth below this very wet soil, because very wet soil has a low bearing capacity and is highly compressible. In addition, should the ponded soil ever dry out, the foundation will be subjected to shrinkage distortion, which can be just as damaging as the distortion that results from heaving soil. Thus, this method is not recommended for reducing post-construction heave.

Should the soil be treated with a stabilization method before the foundation is constructed?

Chemical stabilization has been used effectively to stabilize highways, airport pavements, and large industrial sites. It has also been used successfully on smaller projects, such as single-lot residences and other small buildings, but it is more expensive on a per unit basis than for the larger projects.

There are two methods commonly used in chemically treating expansive soils on small lots. One method is to treat the top 12 to 24 in. (30 to 61 cm) by mixing lime, cement, or fly ash into the soil and then recompacting the mixed soil. This mixing operation requires special equipment which is difficult to operate and maneuver on small lots; hand methods of mixing are not as effective as the mechanized mixing methods. This method also requires laboratory testing to determine the appropriate amount of chemical to mix with the soil to produce the desired effect. However, these chemicals, when used in the proper proportions, when properly mixed, and when properly applied to the soil, are known to effectively reduce the amount of shrink and heave the soil might experience if not treated. It is recommended that a registered professional geotechnical engineer be retained to design and supervise this operation.

The second commonly used method of chemically treating expansive soils is to inject a pressurized slurry of water and lime, cement, fly ash, or potassium chloride to depths of several feet below the ground. The concept is that the pressure will make the slurry flow through cracks in the expansive soil and effectively seal the soil to a specified depth from subsequent penetration of water. The slurry is also expected to interact with the clay particles and reduce the affinity for attracting free water in the same manner as the surface mixing method does. Pressure-injected slurry applications have been quite successful in many instances; however, in many other instances, the injected slurry had no effect on the resulting soil movement. Thus, like the surface mixing method, it is recommended that a registered professional geotechnical engineer be retained to design and supervise the pressure injection operation.

What is a moisture barrier?

A *moisture barrier* is a structure or material that prevents or retards moisture from moving into or out of the soil. Moisture barriers may be vertical, horizontal, or a combination of horizontal and vertical (Figure 2-6). The principle behind a vertical moisture barrier is that if it is attached to a foundation, the distance that water must travel to either get beneath a foundation from the outside or to get out from under the foundation is greatly increased. Because the soil permeability or conductivity is so small, it is hoped that the increased distance—which means an increased travel time—will not permit the soil water content to change appreciably from one season to the next, and the magnitude of shrink or heave correspondingly becomes but a nominal amount.

A similar principle applies to the horizontal moisture barrier (Figure 2-7), except that only a very wide horizontal barrier of several feet will produce the same time of travel effect that the vertical moisture barrier produces. The principal advantage of the horizontal moisture barrier is that it effectively moves the edge moisture variation distance out from under the structure to where it is acting under the horizontal moisture barrier—and where there is less concern if the soil under the moisture barrier heaves and shrinks.

The combination vertical and horizontal barrier might be used if it is difficult to excavate deep enough to place a deep vertical moisture barrier, or if there are lateral constraints that prevent a full horizontal barrier from being used.

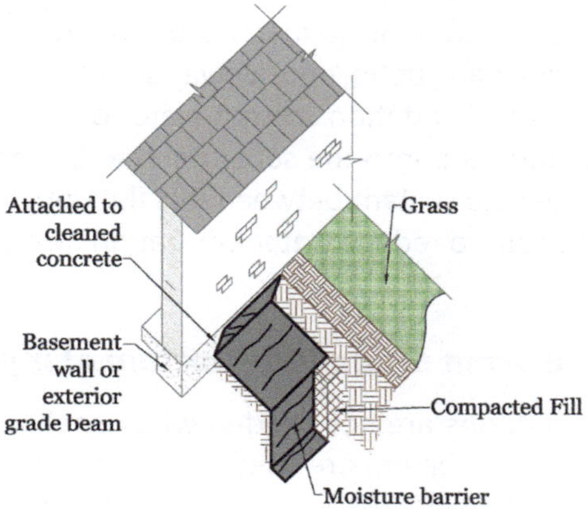

Figure 2-6. Combination of vertical/horizontal moisture barrier.
Source: AccuTech Consultants, LLC, used with permission.

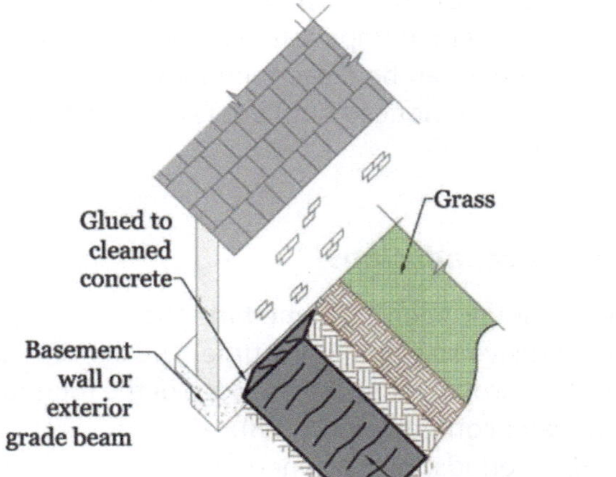

Figure 2-7. Horizontal moisture barrier.
Source: AccuTech Consultants, LLC, used with permission.

Should I use a moisture barrier around my house or building?

Many people have successfully used moisture barriers. However, it is a major construction task and usually is beyond the scope of being installed by "do-it-yourselfers." A common application of the horizontal moisture barrier is to build a concrete sidewalk around the perimeter of the house; large patios and driveways constructed right against the foundation accomplish the same effect. If used, the sidewalk, patio, or driveway should be constructed sloping slightly away from the foundation so that water does not "pond" adjacent to the foundation and seep into the subsurface soil. Proper installation and maintenance of a flexible sealant between this flatwork and the house foundation is also important to reduce water infiltration next to the structure.

Should I install a drain around the bottom of my foundation?

Usually, foundation drains are not needed when shallow foundations (e.g., slabs-on-grade, post-and-beam) are used. There are exceptions, particularly when surface drainage or lot cross-drainage creates problems, then foundation drains for shallow foundations have been beneficial to controlling the movement of the supporting expansive soil.

Foundation drains are more commonly used in association with basements to ensure that the basement backfill is drained or to ensure that water does not enter the basement due to high or rising groundwater tables. It is recommended that a registered professional geotechnical engineer be retained to design this type of drain.

When using foundation drains, it is best if the system uses gravity drainage versus a sump pump system. Pumps can malfunction, and reliance on automatic systems sometimes can be disappointing when the "automatic" part of the system fails and is unnoticed until doors and windows in the building begin to stick or not close.

Should I water my foundation?

The term "water my foundation" is something that is used by persons who have lived with expansive soils for a long time. The concept is to try to keep the soil water content around or under the edge of the foundation at approximately the same water content (Figure 2-8).

Thus, during dry periods, homeowners in the arid or semiarid regions of the country would regularly spray water adjacent to the foundation as a normal part of the lawn-watering regimen. The practice yields surprisingly good results but is applicable only for shallow foundations such as slabs-on-grade. However, the key to watering your foundation is to begin watering the ground adjacent to the foundation early in the dry season. Waiting to begin the watering operation after wide cracks in the soil adjacent to the foundation are noticed may produce the opposite effect of controlling the soil movement; the water may run down the crack and pond beneath the foundation and subsequently produce heaving when there should be shrinkage (without the watering). Thus, "watering your foundation" is not a bad practice and may even produce desirable results if you start early, do it regularly but not excessively, and do not add water to soil crack next to the foundation. Add water to the soil approximately 18 in. (0.5 m) away from the foundation.

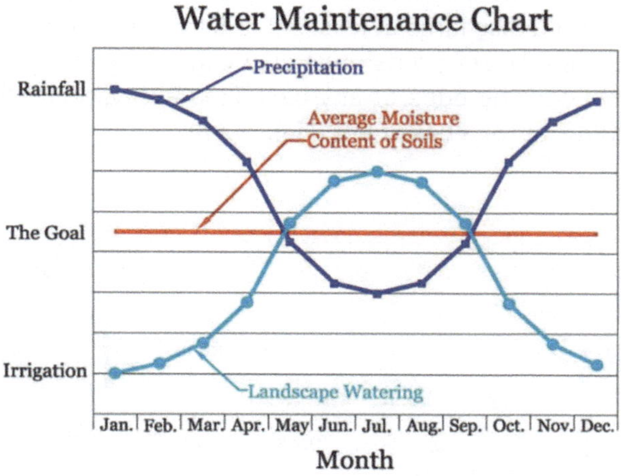

Figure 2-8. Water maintenance.

Source: AccuTech Consultants, LLC, used with permission.

What is meant by positive drainage?

The term *positive drainage* means that the surface of the ground slopes away from the house foundation. Positive drainage is essential to the successful performance of the structure constructed over expansive soils. Owners should, at a minimum, annually inspect the ground over a distance of at least 5 ft (1.5 m) from the perimeter of the building immediately following a rainstorm to see if there are any locations where water is trapped or is ponding. Water that is trapped or ponded cannot drain away from the foundation. Instead, it will soak into the ground resulting in heave beneath the perimeter foundation, if it is a shallow foundation, or perhaps infiltrate into the backfill behind the basement wall if there is not an impervious layer covering the backfill. If any depressions or ponded areas are discovered, they should be filled with clayey soil, properly tamped or compacted, and sloped so that water will flow away from the building. Positive drainage is one of the most important things that an owner can do to avoid or lessen the damage that expansive soils might cause the building.

Should I be concerned about where my downspout discharges?

Downspouts can discharge considerable amounts of water if the roof area is large and the rainfall significant. If the ground surface does not slope away from the foundation, the discharged water will simply pond against the foundation until it soaks into the ground, resulting in heaving beneath the perimeter of the house or building, or it might infiltrate into the backfill behind the basement walls. Most experts recommend that downspouts be directed away from the house and discharge the water at least 5 ft (1.5 m) away from the perimeter of the house into a well-drained area if there is no basement. Downspouts should discharge at least 5 ft (1.5 m) away from the edge of the house or structure. Downspouts must also carry water over and discharge several feet beyond the edge of any backfilled excavation adjacent to the house or building, as for a basement (Figure 2-9).

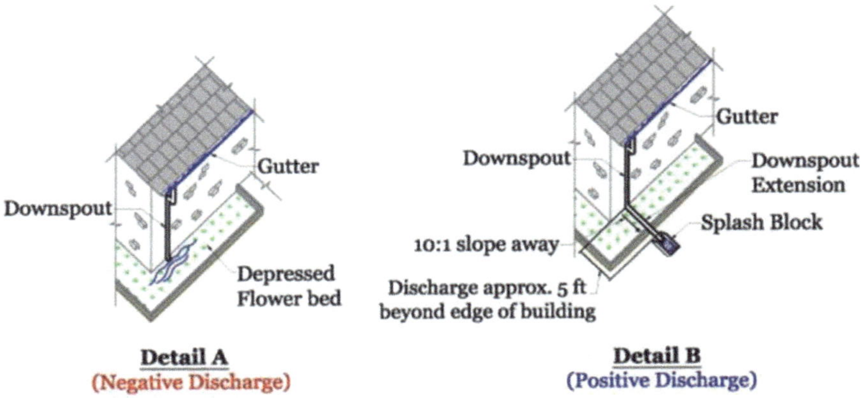

Figure 2-9. Poor and recommended methods of discharging roof runoff water.

Source: AccuTech Consultants, LLC, used with permission.

What can be done when constructing my house that will reduce any damage that might subsequently occur from foundation movement?

Many of the things discussed previously can be used in the design and construction of new houses or buildings that can prove to be beneficial to the long-term acceptable performance of the structure. However, one of the first things that must be decided is how much movement and the effects of the movement (e.g., cracks in the wall) can you tolerate? Thus, a stiffer or stronger foundation, particularly with respect to slab foundations, will permit less deflection or distortion in the superstructure of the building which, in turn, will reflect in fewer cracks. However, stiffer foundations cost more money.

Large shrubs or bushes and especially trees should not be planted close to the house. Smaller bushes or flower beds adjacent to the house or building should not be watered by ponding water in the bed where the bushes or flowers are growing (Figure 2-10). Trees should be planted so that the drip line of the tree at maturity is still several feet from the edge of the house. If the location of the mature tree's drip line cannot be determined in advance, then plant the tree a distance from the house equal to the mature height of the tree.

Drainage Tip

Downspouts can discharge large amounts of water, especially if the roof is large and the rainfall amount is large. If the ground surface does not slope away from the foundation, the water will simply pond against the foundation until it soaks into the ground. This can result in heaving beneath the perimeter of the house, and the formation of cracks, particularly near the drainpipe.

Figure 2-10. Do not excavate a border for flower beds next to houses to prevent ponding of water near the structure.

Source: Marshall B. Addison, Ph.D., P.E., used with permission.

Landscaping Tip

Tree roots will draw moisture from the soil. The reason for planting vegetation away from the house is to limit changes in moisture near the foundation. Two common methods to limit potential damage are to ensure that the tree's mature drip line will be outside the foundation (the tree will not overhang the structure) or a more conservative rule of thumb in which the tree is placed a distance away from the foundation equal to the mature height of the tree.

Added porches, steps, sidewalks, patios, and driveways should not be rigidly connected to the house or building. These minor structures may move differentially with respect to the main structure and may result in damage.

Make sure that sidewalks, patios, or driveways constructed adjacent to the foundation slope slightly away from the foundation so that water cannot pond or pool adjacent to or near the foundation. If a swale was constructed across your property to carry surface runoff water from lots at higher elevations to a stormwater sewer or channel, do not alter or change the swale. Also, ensure that the swales are clean and clear of debris. Similarly, make sure that debris or other material has not accumulated in any gutters and downspouts on the buildings.

Maintaining relative constant soil water content is a very important owner task in mitigating or reducing soil shrinking and heaving. This means watering more during hot, dry periods, but it also means continuing to water during cooler, wetter periods, albeit at a reduced rate.

Your architect and/or engineer can discuss architectural and structural details with you that can help to reduce the effects of shrinking and swelling soil.

How do concrete driveways and sidewalks affect the performance of my house?

Flatwork, such as driveways, sidewalks, and patios, provide several positive influences on the performance of your house. If they are constructed abutting the building's foundation, they serve to prevent the soil around the edge of the foundation from either wetting up or drying out (Figure 2-11a). Driveways and sidewalks can also have a negative influence on a building's performance. If they are cracked, and especially if they have depressions in their surfaces, they can trap water (Figure 2-11b) and/or permit it to infiltrate beneath the cover and result in soil heaving. It is important that driveways and sidewalks slope away from the building, rather than slope toward it (Figures 2-11 and 2-12); if the surface slopes toward the building, rainfall water can pond against the foundation and eventually soak into the ground beneath the paving and result in increased heaving beneath the principal building. Thus, sidewalks, driveways, and patios should be observed at least annually following a heavy rain to ensure that they do not trap water and that they do not drain toward the house and allow water to pond against the building foundation. If either of these situations is discovered, remedial steps should be taken. Proper installation and maintenance of a flexible sealant between flatwork and the house foundation is also important to reduce water infiltration next to the structure.

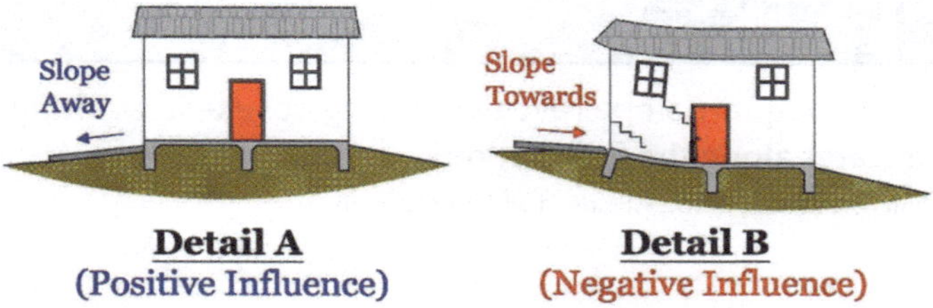

Figure 2-11. Influence of concrete flatwork next to the foundation: (a) good, and (b) poor.

Source: AccuTech Consultants, LLC, used with permission.

Landscaping Tip

To minimize damage from shrink-swell soils, the overall goal is to keep the moisture level below and near your home's foundation at a uniform (constant) level. In general, it is the variation in moisture that causes cracking, distress, or damage. Anything related to water plays a role—how much you water your lawn and plants, if you water them evenly, how your roof is drained, if your water pipes are leaky, if the ground surface allows water to pond near your house, if it has been excessively rainy, or if there has been a drought.

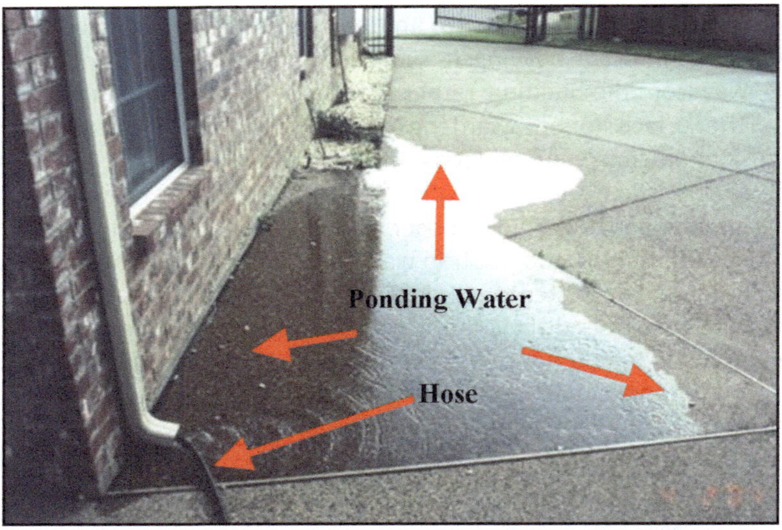

Figure 2-12. Inadequate slope revealed during hose test resulting in ponding of water along the foundation.

Source: Marshall B. Addison, Ph.D., P.E., used with permission.

Why are wingwalls discouraged in houses built over expansive soils?

Wingwalls are little more than decorative extensions of an exterior wall of a house or building. Their usual function is to make the structure seem wider or longer, or they serve some architectural appearance purpose. Seldom do they serve any structural purpose. Because of their light weight and minimal loading on the underlying soil, wingwalls will typically move differentially with respect to the structure to which they either attach or abut (Figure 2-13). They can present an unsightly appearance. If wingwalls must be used, the point at which they meet the principal structure should be *articulated* (i.e., a planned joint).

Should I be concerned about where my air conditioning condenser unit is placed?

The air conditioning equipment that often is placed outside the house is usually the condenser. This is the device that removes humidity from the refrigerated air. The removed water is allowed to drip from the condenser unit, usually onto the ground outside the unit. Care should be taken to ensure that the drain from which the collected condensation water drips is directed away from the foundation, preferably allowed to run across the concrete pad on which the condenser unit is residing and away from the building. Allowing the water to drip between the condenser unit and the foundation, or next to the foundation (Figure 2-14), is inviting the soil in that region to become wetter than the surrounding soil and could result in differential heaving of the foundation in the vicinity of the condenser.

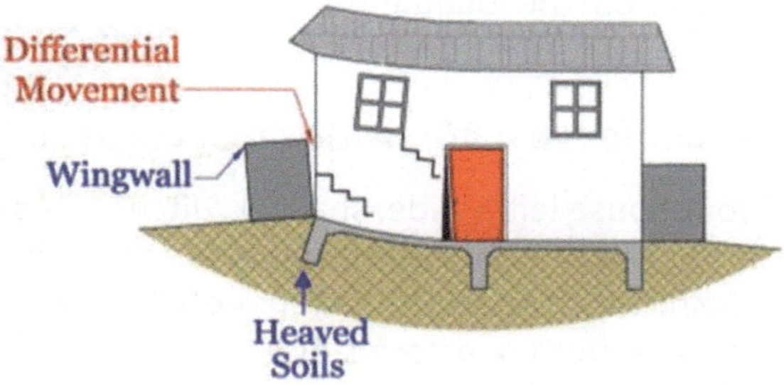

Figure 2-13 Wingwall differential movement.
Source: AccuTech Consultants, LLC, used with permission.

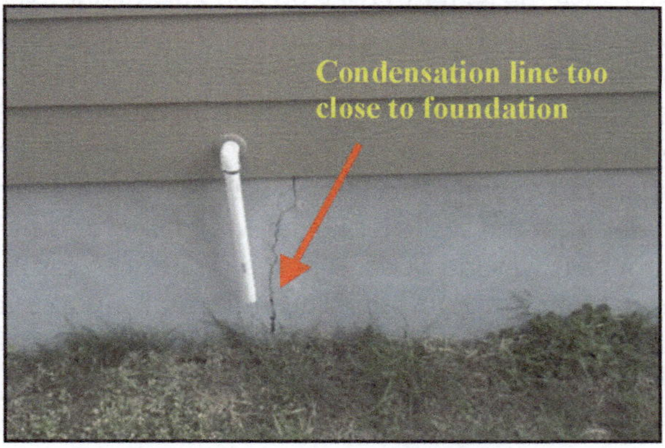

Figure 2-14. Improperly draining air conditioner condensation line.
Source: AccuTech Consultants, LLC, used with permission.

Does it matter where the control valves to my landscape sprinkler system are located?

The valves that control the landscape sprinkler system are notorious for leaking. Because the valves are either covered or at the bottom of deep, narrow pipe, it is difficult for the homeowner to know when the valves are leaking. A common, but unfortunate, practice is to locate the valves close to the house, next to the foundation, or behind bushes which are planted near the foundation (so that they are "hidden" from view but still convenient to the homeowner). If the valves begin leaking, the water can induce heaving beneath the foundation. Locating the valves adjacent to a sidewalk or a driveway may result in wetting of the soil beneath the walk or driveway, possibly resulting in distorted, heaved, and broken sidewalk and driveway pavements. Thus, it is better, albeit often more inconvenient or unsightly, to locate the valves at least 5 ft (1.5 m) away from the foundation.

Why did my builder leave a big, wide ditch across my backyard?

If the builder of your house left a wide, shallow "ditch" across your lot, your lot probably is in the line of drainage from lots further uphill. The purpose of the drainage ditch, which is also called a *swale* (or trench), is to collect the surface runoff water from rainstorms or excessive irrigation and carry it away from all the lots to a drainage channel or a storm drain. It is important that you do not fill in the swale! If you alter the swale, you are likely to invite the runoff water to move toward your house and soak the soil around your foundation, which will result in swelling of the soil beneath the foundation. You might also cause the water to back up and affect your neighbor's house or property, which will likely invite a dispute. If the swale is totally objectionable to you, then you should discuss alternative solutions with your builder or developer. But whatever you do, do not take matters into your own hands and fill in the ditch!

You should also frequently check the swale to make sure that it is clean and free from debris. If the swale is choked with debris or other materials, it will not perform as it was intended. Blocked swales can result in the water finding a new path across your property, and usually in a route where you least want it to go! So, make sure that the swale is clean and ready to carry away stormwater runoff.

Are there any special considerations if I want to build a patio in my backyard?

The patio should not block a drainage swale and it must have a positive slope (see preceding question) and drain away from the house or building. Likewise, a paved driveway or sidewalk that abut the house or building should also slope away from the building. Do not ridigly connect the patio (or driveway or sidewalk) to the existing building's foundation. The two structures will move differentially as the moisture content in the soil beneath each of the structures' changes. If the patio (or driveway or sidewalk) is rigidly attached to the building's foundation, it is likely that the patio will crack and break within a few inches of the end of the device (e.g., a steel rod or dowel) that was used to attach the patio to the foundation.

What special precautions should I take if I build on a hillside or on a steeply sloping lot?

Building on hillsides or steeply sloping lots can present some unique challenges as well as some potentially damaging movements for your house. When building on a slope or steeply sloping hillside, you should retain a geotechnical engineer to provide design and construction recommendations to your architect or builder—and they should be followed!

Are there any special considerations when adding porches or steps?

Yes, porches and steps require special considerations. Because both types of structures weigh less than the house, if the underlying soil experiences volume change, porches and steps will move differentially with respect to the house that they serve. If the porch or steps are rigidly attached to the house, experience has shown that porches or steps will break just beyond the ends of the dowels or other method of attachment. This will likely create a safety hazard and be unsightly.

About Owner Maintenance

How far from my house should trees be planted?

If it is possible to estimate where the tree's drip line will be at maturity, the tree should be planted where at maturity the drip line will be a few feet further from the building. If the house is new and the tree small, the tree will look like it is planted too far away from the house, but the tree will grow! The reason for planting the tree away from the house is because in times of severe drought, the tree's roots will grow further outside the drip line seeking additional moisture. If it is not possible to estimate where the tree's drip line will be, the conservative rule of thumb is to plant the tree a distance away from the foundation equal to the mature height of the tree (Figure 2-15). The Royal Botanic Gardens of the United Kingdom has provided some planting guidelines based on the type of tree (see Table 1-1, page 12).

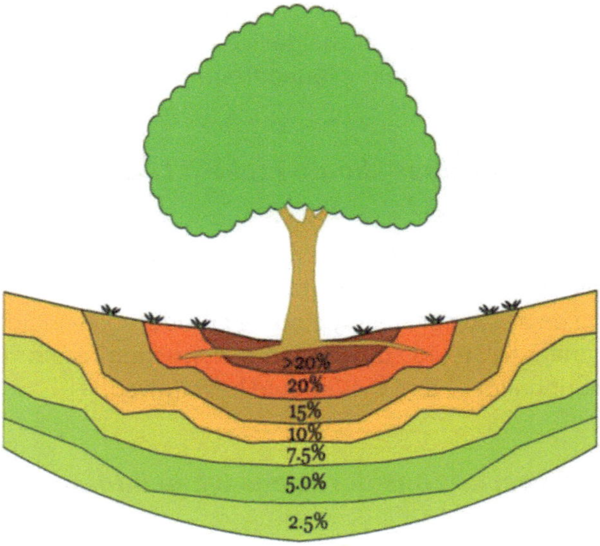

Figure 2-15. Moisture loss in the vicinity of a tree.
Source: AccuTech Consultants, LLC, used with permission.

Should I allow water to pond in my flower beds adjacent to the foundation?

Most definitely not! This is an archaic method of watering vegetation and should be avoided at all costs when the soil beneath the flower bed and the adjacent building is an expansive soil. The practice of flooding flower beds only serves to keep the soil beneath the adjacent foundation wetter than it needs to be and induces excessive heave beneath the foundation.

How important is drainage across my lot and around my foundation?

One of the principal objectives of the owner should be to keep the soil at a relatively constant moisture content. Allowing water to flow adjacent to a building's foundation, or even worse, to allow water to pond against or around the building, will cause the soil moisture to increase and result in increased heaving beneath the foundation. At least annually, the owner should observe the flow of surface water across the lot and adjacent to the house or building. Any locations that indicate that water is being trapped or ponded in a depression—or even if the water is not flowing quickly away from the building—should be modified to improve the drainage condition.

Will leaking water lines or sewer lines under my foundation have any effect on my foundation?

The water in potable water (drinkable water) lines is under pressure. Thus, if a water line should spring a leak, the water will continue to flow from the broken pipe until the water is all gone or until the pressure is reduced to the point where no more flow will occur. An experienced plumber can usually tell if a water line has a leak by running a standard pressure test. If not discovered soon after the leak begins, and if the leak is permitted to continue to allow water to seep into the soil, the result will be localized heaving in the vicinity of the leak.

Expansive soils, however, typically have a very high clay content. As a result, the *soil permeability* or *conductivity* (the measure of the ease or difficulty with which water can travel through soil) is typically low, which means that the water does not move very fast in the soil. As the soil begins to get wet from the infiltrating water, the soil will swell, and its permeability will become even less than it was before the leaking water made it wetter. The low permeability also means that any heaving resulting from the wetted soil will likely not be apparent for some time (perhaps weeks) after the leak first occurs.

Sanitary sewer lines are different from water lines; sewer lines are not pressurized. Sewer lines, except under unusual circumstances, do not have water in them except when a sink, shower, or toilet is draining. A plumber specifically trained in leak detection can determine if the sewer line has a leak by performing a different type of standard pressure test; however, it is difficult to obtain reliable results when performing a pressure test on a nonpressurized line. If the line has a leak, the water level in one of the drains, usually that from a toilet which was removed to perform the test, will usually go down; however, even if the water level drops, the result is often inconclusive and does not necessarily positively determine that the line has a leak. However, even if the line is cracked, the nonpressurized water that normally escapes the sewer line usually has little opportunity to leak out and infiltrate the soil. Thus, a leak in a sewer line typically does not result in severe heaving.

Does it matter how often or how much I irrigate my lawn?

Yes, it does. Ideally, vegetation—trees, bushes, shrubs, lawn grasses, and so forth—should only be watered an amount and at a frequency recommended by lawn specialists in your area. In the case of expansive soils, "more" is not necessarily "better!" Overwatering can result in the soil adjacent to the house getting wetter and causing the soil to heave. On the other hand, it can be equally detrimental to underwater the vegetation, particularly trees and bushes that are close to the building. During times of drought (Figure 2-16), roots from the larger vegetation will seek out additional water to survive and what better place to look for additional water than beneath the foundation of your house or building? Thus, during dry periods or droughts, at least some minimal amount of lawn irrigation should be practiced to prevent the vegetation from removing water from beneath the foundation of your building and causing the soil to shrink. During high demand seasons and droughts, many municipalities impose landscape watering restrictions. Consult with your local water supplier for acceptable methods to water your foundation even when water restrictions are in place.

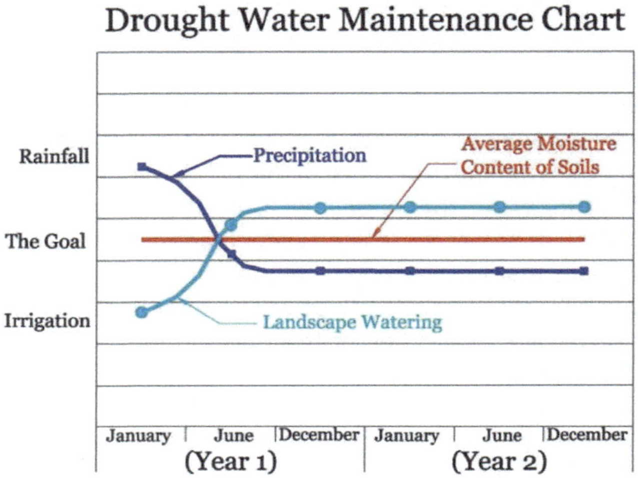

Figure 2-16. Water maintenance during a drought.
Source: AccuTech Consultants, LLC, used with permission.

About Remedial Measures

How does a slab-on-grade foundation work?

A slab foundation is intended to work with the shrinking or heaving soil; it is not intended to prevent the soil from moving. The slab foundation is intended to be strong enough or stiff enough to limit the flexing or distortion that could occur in the superstructure as a result of changing soil volume beneath the slab foundation. To construct a slab foundation so stiff or rigid that it would not permit any movement would be cost prohibitive. If the superstructure is made of brittle materials that are likely to crack with only a little bending, then the foundation must be designed and constructed stiff to limit the bending to less than that which the superstructure can withstand. On the other hand, if the superstructure materials are very flexible, then the foundation can be less stiff and more flexible.

What is tilting?

Surface elevations measured on slab foundations that show a considerable difference from one side to the other with a relatively constant slope across the surface reflect what is called *tilting.* Tilting describes a situation in which one side or end of a building is being raised by heaving soil, while the opposite side or end is not experiencing any movement or is shrinking. Tilting is different from differential deflection or distortion, because a foundation that is tilted does not exhibit bending (Figure 2-17). Thus, a tilted slab may be performing well from a bending viewpoint but, because it is tilted excessively, the structure may not be performing well. Obviously, a slab could experience both bending (distortion) and tilting, but it is important to distinguish between the two concepts.

Distortion refers to how the foundation is being bent or twisted (Figure 2-18). Most damage that occurs to houses constructed over expansive soils results from these conditions.

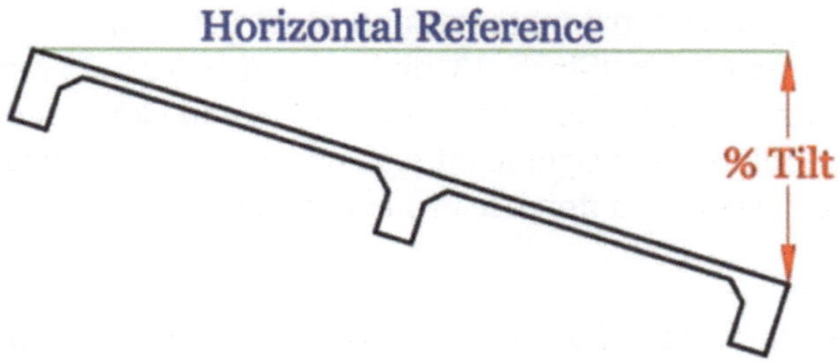

Figure 2-17. Foundation tilt.

Source: AccuTech Consultants, LLC, used with permission

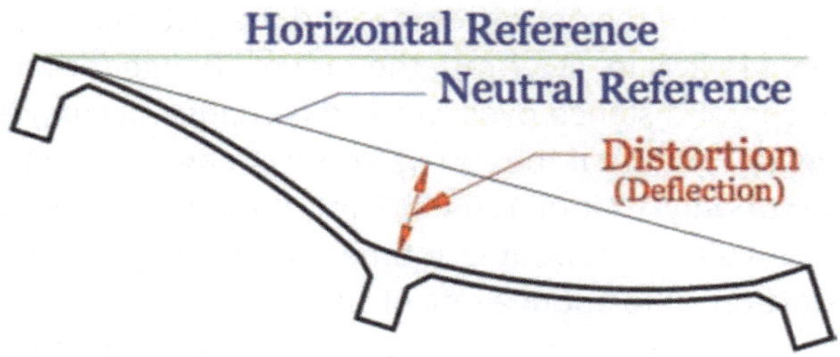

Figure 2-18. Foundation distortion (bending).

Source: AccuTech Consultants, LLC, used with permission.

How can it be determined if a house or foundation is distorted?

A common way to determine if a foundation is distorted is to check the elevation of numerous points on the surface of the foundation. Assuming the foundation surface was a near constant elevation (level surface) at the time it was cast, then the relative difference between the elevations of the various points on the foundation surface will show whether a slab is distorted and the magnitude of the distortion.

If a slab foundation is an inch out of level, is this a slab failure?

This magnitude of differential elevation might produce cracks in the superstructure in one house, but not cause any cracks in another house.

Conscientious builders try to construct the slab foundation with a level surface. They check the elevation of the forms several times as the forms are erected and set. However, studies have shown that most slabs are constructed with more than a half-inch difference between the highest point and the lowest point on the slab. As you might imagine, it is difficult to ensure a perfectly level slab with the materials and construction methods commonly being used. Although a half-inch (13 mm) or even an inch (25 mm) difference between points on the surface of the slab may sound like a lot of error, if the high and low points are a significant distance apart, the superstructure will never even realize that a difference in elevation exists.

When does a difference in floor surface elevation become an important factor?

To engineers, the rate of change of elevation is often more important than the total change in elevation when assessing serviceability or performance of a structure. If, for example, the difference between the high and low points on the surface of a slab is 1 in. (25 mm), and the difference occurs between one side of the slab and the opposite side, and the distance between the two points is 500 in. [41 ft, 8 in. (12.7 m)], the rate of change (or ratio of vertical change to horizontal distance over which the change occurs) is only 1/500 or 0.0020 in./in. (0.2%). But if the 1 in. (25 mm) change in elevation occurred over a distance of only 10 ft or 120 in. (3 m), the rate of change is 1/120 or 0.0083 in./in. (or 0.83%), or about four times greater. Sheetrock can typically withstand distortion up to about a rate of change of 1/300 or 0.0033 in./in. (or 0.33%). Thus, Sheetrock would expect to crack if the assumed 1 in. (25 mm) of differential movement occurred over a distance of only 10 ft (3 m). Because the errors in floor slab elevation that occur as a part of the slab construction are usually very gradual in nature, the fact that a slab is not level at the time of construction is usually not a concern from an engineering viewpoint.

What is structural damage?

Typically, *structural damage* refers to a situation in which a structural member is broken or otherwise damaged to the point at which it can no longer carry or transmit the magnitude of loads or forces that it was intended to carry. Under some conditions, a structural member experiencing an excessive amount of deflection is considered to be damaged.

What is cosmetic damage?

Damage that does not result in structural damage is said to be *cosmetic damage*. Typical examples of cosmetic damage include cracks in Sheetrock walls or exterior brick veneer walls, cracks in concrete floors or basement walls, separating fascia trim, cracks in floor tiles, and so forth (Figure 2-19). Cosmetic damage does not impair the structure's load-carrying capacity.

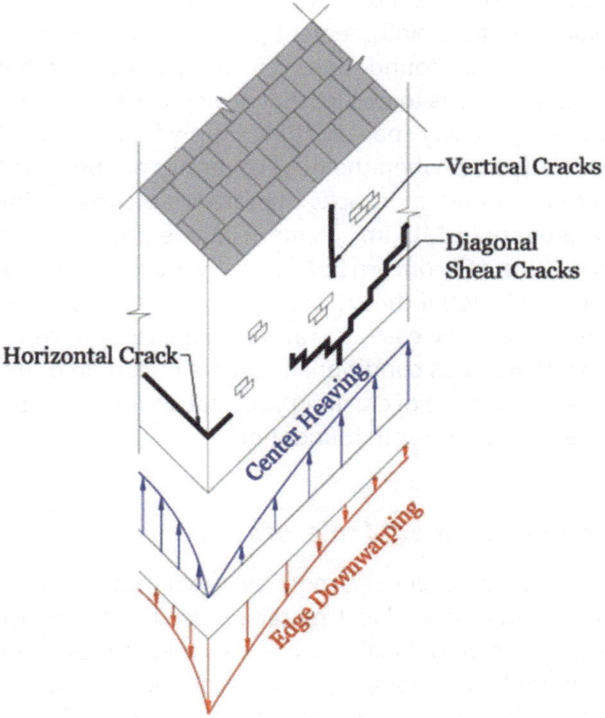

Figure 2-19. Cosmetic damage.

Source: AccuTech Consultants, LLC, used with permission.

How do I know if my house is suffering from expansive soil movement?

Usual symptoms include doors that will not close properly or stick when they are opened, cabinet doors that will not stay shut, diagonal cracks in the wall at the corners of doors and windows (both inside and outside the building), windows that are hard to open and close, and sloping or unlevel floors. Other indicators are gaps between window frames and walls, separations between walls and ceilings, gaps between the garage door and the pavement at either side of the door, and gaps at the corners of fascia trim.

My concrete slab floor has cracks in it. Is this a dangerous situation?

First, it should be recognized that concrete usually cracks within a few days or even hours after it is first placed. Most of the cracks are due to temperature changes in the concrete and shrinkage of the concrete as it dries from its initially plastic state to its more familiar hard condition (Figure 2-20). The principal reason that steel reinforcement (post-tensioned cables or mild steel rebars) is used in concrete foundations is to control cracking that is expected to occur. *Controlling* means to ensure that any cracks that do occur are small in width and usually closely spaced. Thus, simply finding a crack in the concrete slab that is exposed when the floor coverings are being replaced is not necessarily a cause for alarm. Cracks that are wide enough that a common plastic credit card (about 1/16 in. [1.5 mm]) can be slipped into the crack are probably reason for some concern but, typically, the concrete slab is still adequately performing its job if the crack does not exceed about 1/8 in. (3 mm) in width. In virtually every case, a cracked foundation does not indicate a life-threatening, dangerous condition. If you are uncertain or uncomfortable with the size and/or number of cracks in your slab foundation, ask a structural or civil engineer to evaluate your foundation.

How much damage should I tolerate?

This is hard to answer. It seems almost everyone has a different tolerance to cracks and sticking doors. Most materials used in the construction of a house have some degree of flexibility and can tolerate some amount of differential movement before cracking. Stucco exteriors, for example, will crack with only a little movement, particularly if the affected wall is long, whereas hardboard siding will not show any cracking until severe warping and distortion have occurred. Hairline cracking in plaster or Sheetrock interior walls, even if diagonal, may not reflect anything but a change in humidity. If you feel uncomfortable with any of the cracking or other soil movement–related types of damages that you might observe in your house, the best solution is to ask a qualified registered professional engineer to look at your damage. (Not all structural engineers perform damage investigations, but some specialize in such investigations; if the engineer you contact does not routinely perform damage investigations, ask for the name of an engineer who does.)

Figure 2-20. Typical concrete crack.

Source: AccuTech Consultants, LLC, used with permission.

If I notice cracks in my brick outside walls, should I be concerned?

There are typically three reasons that brick veneer (or rock or stucco) walls crack (Figure 2-21): the wall is too long and has cracked as a result of temperature changes causing expansion and contraction; settlement has occurred; or the supporting soil has experienced some volume change. Most temperature-related cracks are oriented vertically. Temperature-induced cracks have nothing to do with soil-related foundation movement. Diagonal cracks or cracks that have the appearance of a stair step have a different cause than vertical temperature cracks. It is often difficult to determine whether exterior stair step cracks are the result of settlement or caused by shrinking or swelling soil. Narrow, or "hairline" cracks that follow the mortar courses are usually not related to significant structural movement and should not be a cause of concern. However, such cracks should be observed on a regular basis to see if the cracks are becoming wider, more severe, or more cracks are occurring.

If some or all of these conditions subsequently occur, you should contact a structural engineer to evaluate your situation.

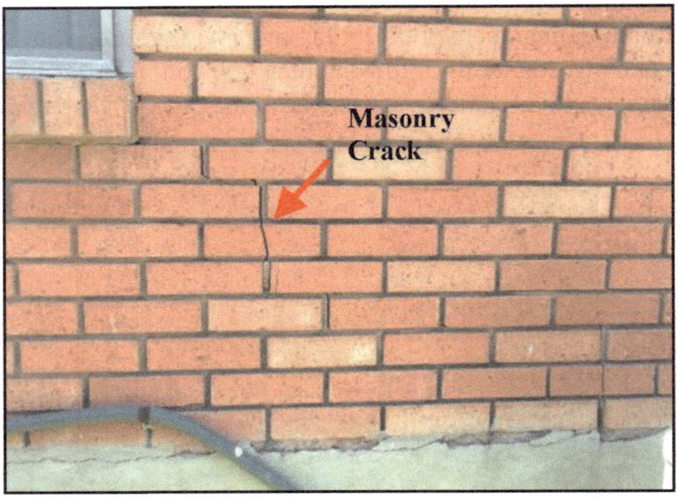

Figure 2-21. Typical brick veneer crack.

Source: AccuTech Consultants, LLC, used with permission.

If I notice cracks in my inside walls, should I be concerned?

It depends. Most American homes and office buildings have Sheetrock interior walls. Sheetrock is installed in pieces that typically measure 4 × 8 ft (1.2 × 2.5 m). The joints between adjacent pieces of Sheetrock are filled with a special material called taping compound. The joints are then covered with paper tape (typically about 2 in. [50 mm] wide and about as thick as two or three pieces of quality copier paper) and then more taping compound until the joint is covered and the joint area is smooth. In a less humid environment, such as inside a house during the winter, just enough shrinkage in the wood framing or trim and other materials can occur to cause a horizontal or vertical crack to occur through a joint between adjacent pieces of Sheetrock.

What actions can I, as owner of the house or building, take to correct expansive soil problems?

Several actions can be taken to reduce foundation movement due to expansive soils. One of the most important actions is to ensure that the soil moisture condition remains relatively constant. Thus, a well-planned watering program for your lawn and vegetation is a simple but important way of controlling differential movement in your foundation. A "good" water program does not necessarily mean that heavy, frequent watering is the solution to controlling soil moisture conditions; a good watering program is one that is appropriate for your location, climate, and vegetation. Local plant nursery experts or the local county agent can provide you with the watering requirements that you need.

The owner should also avoid watering shrubs, bushes, or flowers planted close to the foundation by "flooding" the planting bed. Trees should not be planted close (within a distance inside the mature dripline of the tree) to the perimeter of the building. Positive water flow away from the building foundation should be maintained to ensure that water readily flows away from all sides of the building and does not pond adjacent to or against the building. If you have a basement, ensure that roof runoff water is discharged through downspouts sufficiently distant from the building that the water will not infiltrate through the backfill placed outside the finished basement wall.

Diagonal cracks are usually, but not always, more indicative of superstructure movement or distortion.

©JSN

Engineer's Tip

Often the degree or magnitude of movement is what determines if distress or damage is considered superficial (cosmetic) or structural in nature. If movements are so large that posts, beams, floors, ceilings, stairs, doors, and windows do not connect or work properly, it is time to have a professional engineer evaluate the situation. In some circumstances, homes may become unsafe. If movements are small and the distress is limited to small gaps, cracks, or windows and doors that stick slightly, but are still operational, this damage is usually considered cosmetic. It is also possible that if the foundation soil moisture changes back to a previous condition that these types of distress may lessen or "heal" with time.

How can a professional assist me in correcting or repairing expansive soil damage to my house or building?

If your house or building has experienced excessive distortion, the foundation may have to be *underpinned.* Underpinning is a commonly used repair method and is a process whereby the foundation is modified or repaired and, in doing so, the foundation is partially releveled. Usually, underpinning repairs involve installing piling or drilled piers beneath the foundation to give it additional support. Underpinning repairs should be done only by reputable foundation repair companies and should include an analysis by a registered professional engineer (civil, structural, architectural, and/or geotechnical) regarding the location, the type and extent of underpinning or foundation repair, and the techniques or methods to be followed in the underpinning or repair process. The engineer's evaluation will also consider such things as identification of causative factors of foundation motions, potential for future shrink-swell motions, extent and type of cosmetic and structural damages, the age and value of the house, the applicable repair options, the cost of the repairs, and the impact of the type or method of repairs on the habitability of the building during the repair period. The engineer should supervise the repair process.

Horizontal and/or vertical moisture barriers, which reduce the amount of movement that the building experiences along the perimeter of the foundation, can be installed. In some instances, injection stabilization procedures have been successfully used. If the foundation distortion severely damaged the footings, basement wall, or the slab foundation, the cracked concrete may have to be rejoined using high-strength epoxy compounds injected under pressure into the cracks. Do not be upset at the engineer if he or she recommends waiting several months or even more than a year to have repairs made; the engineer is delaying the repairs to permit monitoring of the heaving and shrinking, which may stop, which will permit the repairs to be more effective and long term.

After remedial actions have been taken, how long will it take for cracks to close?

Some cracks may never close, even after the foundation has been repaired. If you are concerned about these types of cracks, you should call the engineer that supervised the repairs. Most likely, the engineer has determined that the still-open cracks offer no current or future decrease in the structural performance of the building, and the expense required to close or seal the cracks was unnecessary. However, if you have any questions about cracks remaining after repairs have been completed, do not hesitate to address your concerns or questions to the engineer. Also, it is not uncommon for repaired cosmetic damage to reopen or recrack at some date after repairs have been completed. Normally, this is not a cause for alarm; it only means that the plane of weakness in the finishing material that was caused by the original distortion will never be as strong as it originally was and can be caused to reopen by such simple things as a change in humidity from one season to another. If you have concern about the number, arrangement, or severity of the reopened cracks, do not hesitate to contact your engineer.

Can I expect cracks to appear again after remedial measures have been taken?

It is likely that hairline cracks will appear again following repairs. These cracks might occur in the building's exterior, its interior, or in its foundation. If the cracks are small, the cracking is likely due to the structure simply redistributing the loads after the repairs. Remember, the building is a little different than before it was repaired, and the building needs to redistribute the loads and forces acting on it in a little different manner than it did before it was repaired. This process takes some time. Again, if cracks occur in places that you do not think cracks should be appearing or if the cracks that appear are larger than you think they should be, ask your engineer to look at the cracks.

Conclusion

This book has described what makes expansive soils "expansive." It has also described many of the things that cause expansive soils to shrink or swell and how those causes can be prevented or controlled. Many questions that homeowners, builders, and realtors often face have been answered. Obviously, some special situations have not been addressed in these short discussions, but if the reader still has questions, many sources are available to provide additional, more specific, or more technical information, including ASCE, consulting engineers specializing in geotechnical engineering, and civil engineering departments within local universities.

References

ASCE. 2016. *Living with shrink/swell soils: The homeowners' guide to maintenance of shallow foundations*. Reston, VA: ASCE Shallow Foundations Committee Geo Institute. https://www.geoinstitute.org/sites/default/files/inline-files/LivingWithShrinkSwellSoils.pdf.

Building Research Establishment Digest. 1985. *The influence of trees on house foundations in clay soils*. Vol. 298. Bracknell, England: Bracknell Publishing.

Dixon, J. B., and S. B. Weed, eds. 1977. *Minerals in soil environment*. Madison, WI: Soil Science Society of America.

Grim, R. E. 1968. *Clay mineralogy*. New York: McGraw-Hill.

Holtz, W. G., and S. S. Hart. 1978. "SP-11 Information for the Colorado homeowner: Home construction on shrinking and swelling soils." *Soil. Special Publication*. Denver, CO: Colorado Geological Survey, Department of Natural Resources.

Jochim, C. L. 1987. "SP-14 Home Landscaping and maintenance on swelling soil." *Swelling Soil. Special Publication*. Denver, CO: Colorado Geological Survey, Department of Natural Resources.

Jones, D. E., and W. G. Holtz. 1973. "Expansive soils: The hidden disaster." *Civil Eng*. 43(8): 49–51.

Krohn, J. P., and J. E. Slosson. 1980. "Assessment of expansive soils in the United States," *Proc., 4th Int. Conf. on Expansive Soils*, 596–608. New York: ASCE.

PTI (Post-Tensioning Institute). Committee DC-10. 2018. *Guide for performance evaluation of SOG foundations*. Farmington Hills, MI: PTI. https://www.post-tensioning.org/publications/store/productdetail.aspx?ItemID=DC108&Language=English&Units=US_Units.

Residential Foundation Evaluation and Repair Subcommittee 2008. "Guidelines for the evaluation and repair of residential foundations." Texas Section: ASCE. http://www.geotecheng.com/papers/guidelines/asce_residential_repair_guidelines.pdf.

Wray, W. K. 1989. "Mitigation of damage to structures supported on expansive soils." *Final Rep., Vol. I*. Washington, DC: National Science Foundation.

Index

D

E

F

G

R

S